W. H. Bragg

Studies in Radioactivity

Verone

W. H. Bragg

Studies in Radioactivity

1st Edition | ISBN: 978-9-92500-071-5

Place of Publication: Nikosia, Cyprus

Erscheinungsjahr: 2016

TP Verone Publishing House Ltd.

Reprint of the original, first published in 1912.

W. H. Bragg

Studies in Radioactivity

Verone

PREFACE

In this book I have dealt mainly with the phenomena attending the passage of the α, β, γ and X rays through matter. It is a subject which I have myself tried to investigate: and the book was undertaken in response to a request that I should write a short account of my experiments and of the conclusions to which they had led. My own work has been, however, so bound up with and dependent upon that of others that it could not be considered by itself with any profit. I have therefore confined the book within the limits, not of my own enquiries, but rather of the field in which the enquiries have been made. Though this is a limited portion of the domain of radioactivity, it has been difficult to consider it in any detail in the space allowed. I have tried only to give an account of the principal features of the special subject, and to draw together the things which most invite comparison. The statement is far from complete in a historical sense, and the reader who would learn more details of the work which has been done should consult the original memoirs, or books which can really claim to be treatises on radioactivity, such as those of Rutherford or Mme. Curie.

Certain general conclusions seem to me to follow from a comparative study of the three types of radiation, and I hope that I have made this appear from what I have written.

In the first place, it is interesting to observe the absence of any evidence of true secondary radiation; that is to say, of an ionising radiation which derives its energy from matter under the prompting of primary rays. That to which the name has been provisionally given draws its energy from the primary alone, and we can, up to the present, claim no power of causing the atoms to unlock and distribute any stores of energy which they may possess. This is not, of course, the first occasion on which the conclusion has been stated.

Again, it is remarkable that there should be so little evidence of the influence of molecular association upon radioactive phenomena. When an atom acts upon a passing α or β or γ ray, it is unsupported by any other atom, even of those belonging to the same molecule.

It is impossible to avoid being struck by the strong family likeness which the three types of radiation, α, β and X or γ, rays, bear to each other. The α rays are positively charged, the β rays negatively, the X and γ rays are uninfluenced by electric and magnetic fields. But, putting aside these differences and their immediate consequences, in their laws of penetration and of scattering, in their actions on matter and the reactions which they suffer themselves, the three forms of radiation differ in degree rather than in kind. If it is assumed that the action of each form is direct and requires no assistance from any other form, it is difficult to believe at the same time that the α and β radiations are corpuscular, and that the X and γ rays are spreading pulses in the æther. The distinction in form is too great: the X and γ rays have corpuscular properties also.

I believe, however, that the assumption is wrong: and that the X and γ rays act only through the intervention of β rays. This is accomplished by means of a complete interchangeability between the X or γ ray on one hand and the moving electron on the other, a change which may be brought about during the passage of the ray or the electron through the atom. This is one of the most striking of the general conclusions to which I have referred. It explains the great bulk of the X ray phenomena with readiness and simplicity, and, moreover, it bids fair to be useful in the still wider field of general radiation. I have tried to show that the interchange must take place with little loss of energy. Papers by R. Whiddington and C. T. R. Wilson, published so recently that I have been unable to refer to them in the book, accentuate still further the reality and importance of the conception and simplify it by showing that the transformations imply no loss of energy at all. Wilson's most recent photographs of the clouds formed on the tracks of ionising agents are far better than those which I have been able to reproduce.

The principle of interchangeability also leads at once to a corpuscular hypothesis of X and γ rays. The corpuscular idea correlates the main facts in a fashion which is convenient both for thought and for experiment. I think it is just to say that the æther pulse idea has been for some time unproductive. It is only by the aid of numerous and very special assumptions that it can be made to account, even to outward seeming, for the phenomena of the scattering and the absorption of X rays and the production of the secondary radiation. It seems to me better to put it aside provisionally and to take the interchangeability of X ray and electron as a new starting-point. From this, fresh opportunities of advance in knowledge open out in all directions, and after all that is the one sufficient justification for any hypothesis. To take such a step is no denial of all connection between X rays and electro-magnetic phenomena: it is but to put down one tool and to take up another better fitted for the moment to the work in hand.

I am glad to take this opportunity of acknowledging

my debt to Prof. R. A. Gregory, the editor of the series to which this book belongs. He has been most kind in allowing me to benefit by his experience. He set me my task, but he has done his best to make it an easy one.

W. H. BRAGG.

CONTENTS

CONTENTS

STUDIES IN RADIOACTIVITY

STUDIES IN RADIOACTIVITY

CHAPTER I

PRELIMINARY EXPERIMENTS

Of the researches described in this volume, those for which I am myself responsible had their origin in the following way. The Australasian Association for the Advancement of Science met in Dunedin, New Zealand, in January, 1904. It fell to my lot to prepare the presidential address in the section dealing with astronomy, mathematics, and physics. At that time the newly discovered facts of radioactivity following on the isolation of the electron were opening up a new and wonderful field of physics, and it seemed that the address might well be devoted to a survey of the positions which had been gained. I therefore endeavoured to review the researches of Lenard on the passage of cathode rays through matter, the work of J. J. Thomson on the electron, and the properties of the new radiations which had been investigated by Becquerel, the Curies, Rutherford and others.

It had become clear that the electron was able to penetrate the atom with ease under suitable conditions: and the work of Lenard showed that most of the phenomena occurring when a stream of projected electrons forced its way through matter could be accounted for by the aid of this discovery, if it were also assumed that the electrons could suffer some sort of "absorption" in the matter penetrated. In the beautiful series of drawings given as an appendix to Lenard's paper in the *Annalen der Physik*, LI,

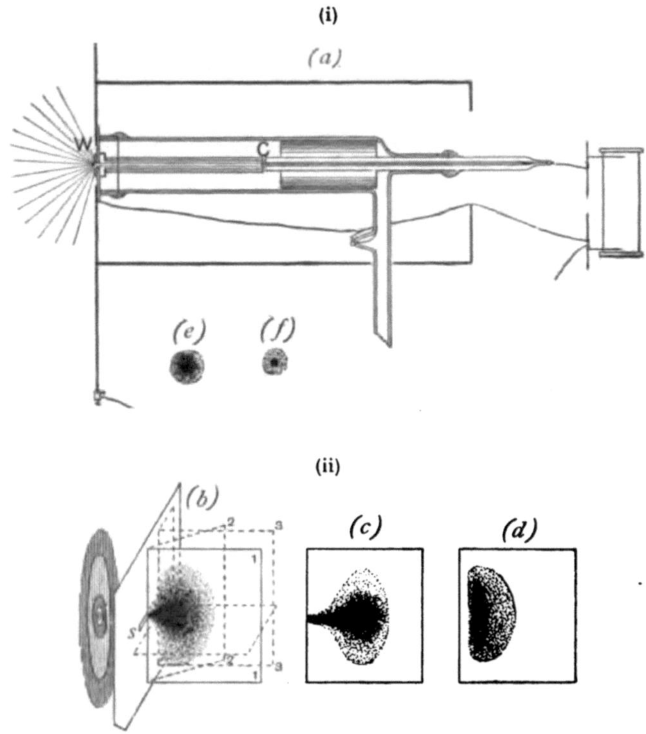

FIG. 1

(i) Lenard's experimental tube. Cathode rays are shown proceeding from the cathode *C* and striking a very thin aluminium window at *W*. Some of them pass through the metal and appear as a diffused bundle of radiation outside, visible through phosphorescent action on the air or on suitable screens.

(ii) The shaded area on the left is the end of the tube, and the small circle in the centre is the aluminium window. A screen with a slit (*s*) is placed as in the figure, and the drawing shows the appearance of the rays after they have passed through the slit. A phosphorescent screen can be placed in the positions 1, 2, or 3. The effects are shown in (*b*), (*c*), and (*d*) respectively. It is clear that the pencil of rays does not spread by a simple widening of its borders, but that the main stream holds on while individual electrons are diverted from it in succession; (*e*) and (*f*) illustrate this still further. They are the effects observed when the screen is put straight across the stream, here moving in an attenuated gas. Figs. (*b*), (*c*), (*d*) are full size; (*e*) and (*f*) are reduced to one-half, (*a*) to one-fifth.

1894, these two principles are clearly illustrated. Lenard considered that the strength of a stream of electrons (Lenard rays) radiating from a source of small dimensions could be represented at a distance r from the origin by the formula $Ae^{-\lambda r}/r^2$. Here A is a constant, the exponential factor implies a loss of energy in passing through matter, and the denominator expresses the spreading from the point. Even at the present time there is no complete explanation of the exponential term, but it is clear that it depends at least in part on the scattering of the electrons by the atoms which they encounter. Lenard's figures show this effect plainly. Whether there is also such a thing as the actual arrest of an electron by an atom while the former is travelling at a high spead, and whether the electron gradually loses energy in its zig-zag flight, and, if both these effects occur, to what extent each influences the absorption, are questions still debated. Lenard shows that the scattering is there, and that it results in the gradual diffusion of a stream of electrons initially moving in one direction, so that the boundaries of the stream become ill-defined and lost.

There was an idea at the time of which I am writing that an atom consisted of an assemblage of electrons separated by empty spaces of dimensions which were great in comparison with the volumes of the electrons themselves. If one electron could thread its way through such a collection, being endowed with sufficient speed to enable it to break through the fields of force into which it must enter, it would follow that a collection of electrons could do the same. An α particle, being an atom and therefore such a collection, should if endowed with enough speed be able to pass through any atom which it met. It now struck me that in such a case the α particle should show no sign of scattering. For even if an electron of the α particle's system should come into collision with an electron of the atom's system, the collision ought to have no appreciable influence on the movement of the α particle, the latter being thousands of times as heavy as the electron. Consequently the α particle should pursue a perfectly rectilinear course, passing without deflection through all the atoms it met. Experimenters on the "absorption" of α radiation by matter had assumed that an exponential law ought to govern the results but—to quote from the Report of the A.A.A.S., Dunedin, 1904—"it cannot be correct to say that the amount of the radiation which penetrates a distance x is proportional to the expression e^{-ax}: it must rather be proper to say that—

"(1) The number of α particles penetrating a given distance does not alter much with that distance until a certain critical value is passed, after which there is a rapid fall.

"(2) The energy of the α particles penetrating a given distance gradually decreases as the distance is increased, and dies out at the same critical value.

"These statements are the expression of what we should expect if ionisation, consuming energy, were alone responsible for the absorption of the radiation. And it appears that this must be an important cause in the case of α rays, since the latter do behave to a marked degree in the way described. In the case of β rays there is something of this effect, showing that the β rays do lose energy on their way: but the absorption of β rays is mainly due to scattering.

". . . There is no secondary radiation from α rays as Becquerel has already shown (*C.R.* cxxxii, p. 371) and this agrees with our idea that the α rays are not stopped by scattering. They are not turned to right and left as they pass through a substance in the manner of β rays."

These ideas received strong support from an experiment described by Mme. Curie (Congrés Inter. de Phys., Paris, 1900, *Rapports*, vol. iii., p. 102). Some radioactive material, polonium, was placed close to the opening of an ionisation chamber. When the material was more than 4 cm. away there was practically no ionisation, but as it was pushed closer and passed a certain critical point, the ionisation current appeared quite suddenly and increased rapidly with diminishing distance. Now if the *a* particle pursued a straight course without deflection, and if it spent energy in producing ionisation along its track, it must have a *range*, and the ionising effect of a stream of *a* rays would end suddenly at a definite distance from the origin of the stream, a distance which would depend on the initial speed of the particle and the nature of the material traversed. Mme. Curie's experiment could be simply explained on the supposition that the range of the *a* particle of polonium was about 4 cm. When a sheet of aluminium 0·01 mm. thick was placed in the way of the *a* rays, the same experiment could be repeated; but Mme. Curie found it necessary to bring the polonium about 2 cm. nearer to the opening in order to show any given phase of the phenomenon. The weight of such a sheet of aluminium is nearly equal to that of a sheet of air 2 cm. thick at ordinary pressure and temperature. It was on evidence and on considerations such as these that the conclusions already given were based.

I had not the means at that time of making any experimental test of the hypothesis which I had been led to construct. But a few months later some radium bromide was placed at my disposal through the generosity of a constant friend of the University of Adelaide, Mr. R. Barr Smith. With the assistance of Mr. R. Kleeman I arranged an appropriate experiment. The first object was to show that the effect of the *a* particles ceased abruptly at a certain distance from their starting-point, and in order to do this the apparatus of Mme. Curie was used in an altered form. The *a* radiation was limited by lead stops

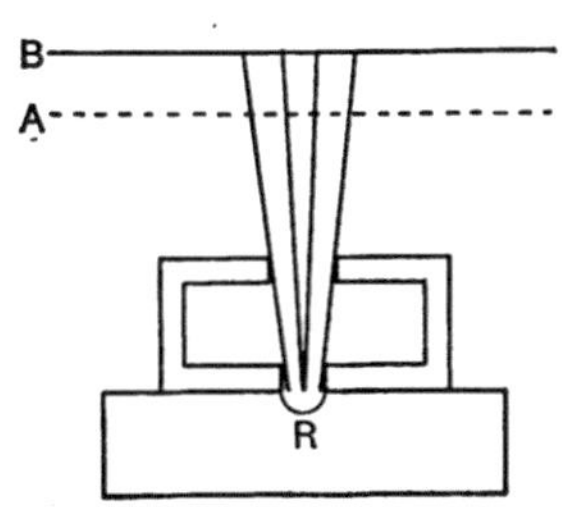

FIG. 2.—*A B* is the ionisation chamber; the upper wall of metal, the lower of metal gauze. The *a* rays stream up from the radium at *R*.

to a narrow pencil, and the ionisation chamber made wide and shallow. With this arrangement the whole of the pencil of α rays would enter the ionisation chamber at the same moment as the radium was brought nearer to it, and with further approach the same pencil would be intercepted and tested at various points. In the original experiment of Mme. Curie closer approach brought about the inclusion of more and more α rays, and interpretation was so much the more complex.

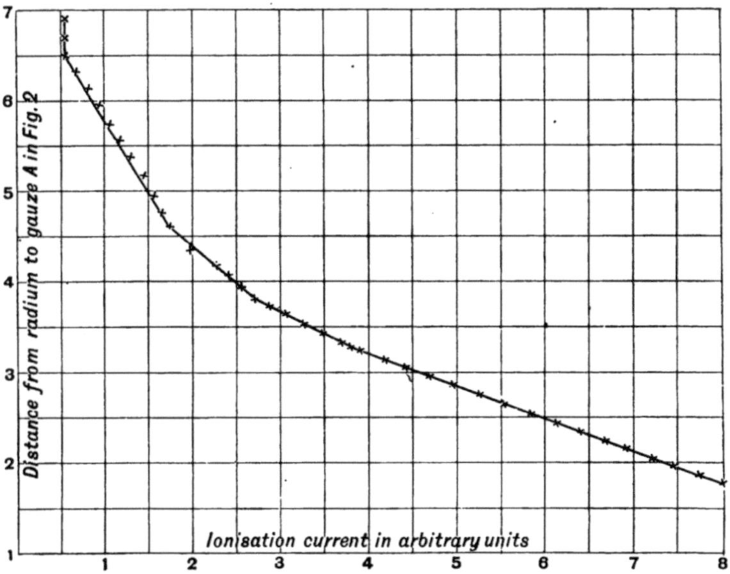

FIG. 3.—Ionisation curve of a thick layer of radium bromide.

The results at once justified the hypothesis, for they showed that while the chamber was yet more than 7 cm. from the radium, there was an effect which altered only slightly as the chamber was moved closer, but that a sudden change set in at that distance and thereafter the ionisation current grew rapidly. When the value of the ionisation current was plotted against the distance, it appeared that the relation between the two was represented nearly by a straight line from about 6·5 cm. to 4 cm., at which point the current began to increase much more rapidly. After turning a corner at this distance the line again became straight, but the inclination to the axis was finally only one-third or one-quarter of what it had been.

An explanation of this result suggested itself at once. The radium layer was of such a depth that α particles from its lower strata were unable to pierce the matter that lay above them: in other words the thickness of the layer was greater than the range of the α rays in radium bromide. Consequently rays with all sorts of velocities would be emerging from the layer, and as the distance between the radium and the chamber diminished the latter came within range of a steadily increasing number of rays. It was only necessary to make one more supposition, *viz.*, that the α particle produced ionisation uniformly along its track, and the straight line feature was explained. As for the existence of two straight lines it suggested that there were α particles of more than one range, and that the distance of 3·5 cm. or thereabouts was the range of α particles two or three times as numerous as the more penetrating stream first observed. Now it was to be remembered that Rutherford's theory of radioactive change presumed the existence of several derivatives of radium, alike in their capacity of emitting α particles, but quite different in their physical and chemical properties. It might therefore be reasonably expected that their α particles would have different initial velocities and ranges. It would accord well with the experimental result if all the particles emitted by any one substance had nearly the same range. Rutherford had proved the existence of four sources of α particles among the substances which would be present in a newly prepared layer of radium, *viz.*, radium itself, the emanation, RaA, and RaC. The substance RaB apparently emitted β rays only, and the other α ray giving products took so long to accumulate that they would not be effective for many years after the preparation of the radium film.

If therefore the hypothesis was right we ought to find α rays of four different ranges, and the curve should be like four consecutive sides of an irregular polygon. A little consideration showed also that the irregularity might well extend to the lengths of the sides only and that the angles might be equal to each other (see p. 19). For, according to Rutherford's theory, any radium preparation which had been in existence for two or three weeks, and had been prevented from losing the "emanation" or any other product, had reached the stage of equilibrium. Radium and its derivatives existed within it in such relative quantities that the number of atoms of each product which formed in a second was equal to the number of atoms of the same product which disappeared in a second and became atoms of the next product succeeding. In consequence the four streams of α particles which resulted from or accompanied the four changes, namely Ra to emanation, emanation to RaA, RaA to RaB and RaC to RaD, were alike numerically. The rough curve already obtained showed at least the first and the fourth slope, that is to say the two extreme sides of the fraction of a polygon, and the irregular curve joining the two might possibly be resolved into the two intermediate sides if the apparatus was made more accurate.

Meanwhile, there was one simple test which could be applied immediately. We dissolved some of our radium bromide in water, so that the emanation was set free and escaped. The solution was evaporated down in a shallow glass vessel and the deposit soon contained very little beyond the one radioactive substance, radium itself; for the emanation was driven off, and the substances derived from it quickly disappeared in the absence of their source. Repeating the range-finding experiment with the layer prepared in this way, we looked for an ionisation curve consisting of one straight line only, due to the α particles of radium itself. I remember that for some unjustifiable reason we expected to find these α particles penetrating the air to the longer range of 7cm., and that when the distance was almost halved and yet no current had appeared in the chamber we were seriously alarmed as to the result of our inexperienced handling of the radium salt. At a distance of 3·5 cm., however, the current set in suddenly, and the experiment was clearly successful. It showed that with the removal of the radium derivatives the long range α particles had disappeared, and that of the four radioactive substances which had been in the layer the radium itself emitted the particles of least energy. The conclusion was in agreement with the hypothesis that each substance had an α particle of definite range, and it only remained to find and allot the remaining three. One range was close to 7 cm.; the other two probably were a little more than 3·5 cm., since the slope of the curve below that value was about a quarter of its initial slope at 7 cm.

At this stage we discovered an effect which was not expected and which simplified the subsequent work. As the piece of glass with its radium deposit was gradually brought still closer to the ionisation chamber the current grew rapidly, but after another 5 or 10 mm. of approach it became stationary for a short space and then to our surprise became less and less. This change in the form of the curve was clearly due to the reduction in the thickness of the radioactive layer which could now be penetrated even by the particles from its lowest stratum. After the approach had been sufficient to take in these particles, no more were to be included by any further diminution of the distance. An explanation had still to be found for the decay of the current as soon as all the α particles had become effective, and the only supposition open was that their efficiency as ionising agents varied along their path and became greater as their speed declined. After all Durack had found a similar effect in the case of β particles. Ionisation might be supposed to result from the presence of the α particle within the atom, and the chance of ionisation of any atom through the intrusion of the α particle within its boundaries might be greater the longer the intrusion lasted. A comet entering the solar system would produce perturbations depending in part on the time it took to cross. I think it may be said that further observation is still favourable to this idea. Penetration must precede ionisation, and the energy which, for example, an electron must possess in order to ionise by collision is not exactly that which is necessary to dislodge a second electron from the atom but rather that which the ionising agent must possess in order to enable it to break through the boundaries of the atom and make its way into the interior, from which it may or may not emerge. The chance of ionisation by the α particle therefore increases as it slows down.

The ionisation curve of a thin layer of radioactive material is quite different from that of a thick one, and is much easier to interpret; for the various ranges are better separated from each other. Using a thick layer, we were unable to improve on the curve shown in Fig. 4, in which the edges of the polygon are not shown with accuracy. The figure is taken from a paper read before the Royal Society of South Australia in September, 1904, and published in England in the following December (*Phil.*

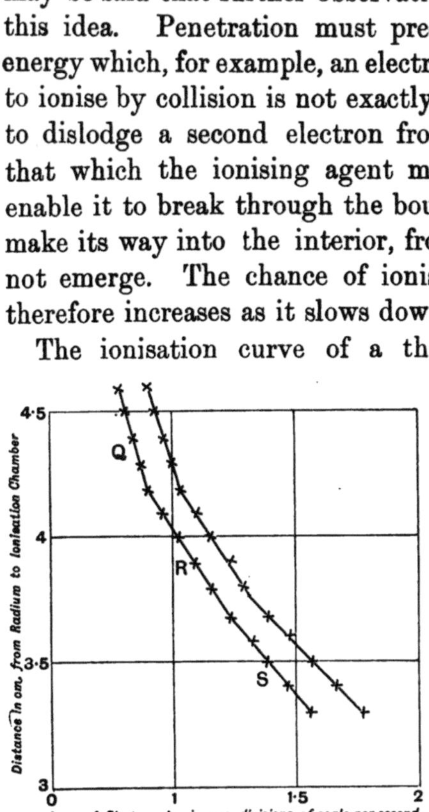

FIG. 4.—Portion of ionisation curve of thick layer of radium bromide, showing approximately the ranges of radium and radium emanation.

Mag., December, 1904). With the thinnest layer we could then prepare, we obtained the curve in Fig. 5. As will be explained, we were afterwards able to obtain much better results (see Fig. 10 on p. 21), principally because we then had the use of a little pure radium bromide which Mr. F. Soddy had most kindly sent me. But the earlier curves showed the ranges of the four sets of *a* particle fairly well, and from curves like that of Fig. 5 we were able to draw the following conclusions.

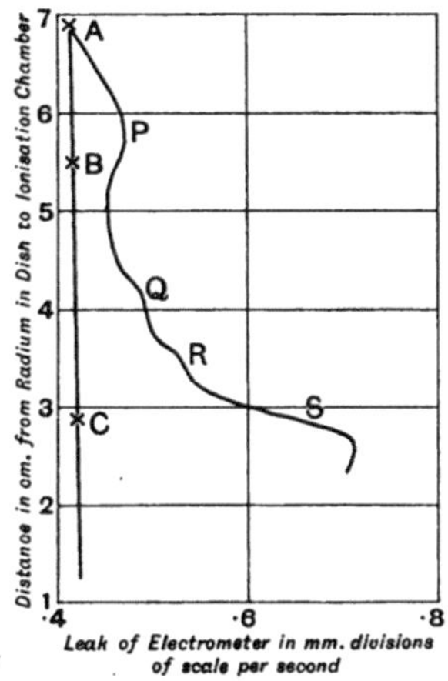

FIG. 5.—Ionisation curve of thin layer of radium bromide. The line *ABC* represents the effects when the *a* rays are cut off by a thin screen.

(*a*) The *a* particle is not appreciably scattered in passing through matter but is "absorbed" only through the expenditure of its energy on ionisation.

(*b*) The *a* particle has a definite range in any given material depending on its initial velocity.

(*c*) Radium in a state of equilibrium contains four substances, each of which ejects *a* particles at the same rate but with different initial velocities.

(*d*) The range of the *a* particle of radium itself is about 3·3 cm.

(*e*) The *a* particle produces more ionisation as its speed diminishes.

We were also able to assign the longest range of about 7 cm. to RaC, and to show that the ranges of the other two substances emitting *a* particles were about 4 cm. and 4·5 cm. respectively. We showed that the ionisation curve could follow by its changes of form the loss of all the radium derivatives through heating the radium or dissolving it in water, and the gradual restoration of these products at their proper rate according to

Rutherford's theory. But the curves which we obtained with improved apparatus and better methods of measurement show these latter results so much more clearly than the original curves that I shall postpone their discussion. I shall first describe the apparatus in the final form which it received in these experiments, and then consider in turn the principal phenomena of the α rays.

CHAPTER II

THE RANGE-FINDING APPARATUS

A DRAWING of the apparatus used in determining the constants of the α rays is given in Fig 6.

The direct object is the measurement of the ionisation due to the α rays at various distances from their origin. The radium is placed in a very thin layer on a platinum plate at *RR*. Over this stands a bundle of about 150 upright tubes *TT* made of thin copper, each a centimetre long and two millimetres in diameter. Only those α particles emerge from such a bundle as leave the plate in an almost vertical direction, all others striking the walls of the tubes and burying themselves therein. The ionisation chamber is the space between the plate *QQ* and the parallel sheet of gauze *gg*. It is three millimetres in depth and its diameter is such that at all distances between the chamber and the plate the whole of the α ray stream enters the chamber. The gauze is carried on three glass pillars of which the one shown in the drawing is perforated along its axis to allow of electrical communication between the gauze and the battery. The pillar which supports the plate *QQ* passes through a glass plate represented by the dotted area, and all joints are made airtight by the use of india-rubber washers. The radium can be raised and lowered by the large screw head at the bottom of the figure; this operates on a long rod which passes through a stuffing-box into the chamber and carries the radium on its upper end. The rod is fitted with an index which travels over a scale graduated in cm. and mm.; a lens being used to read its position. The scale is calibrated when the instrument is put together, and is adjusted so that its readings give the distance between the radium and the middle of the ionisation chamber.

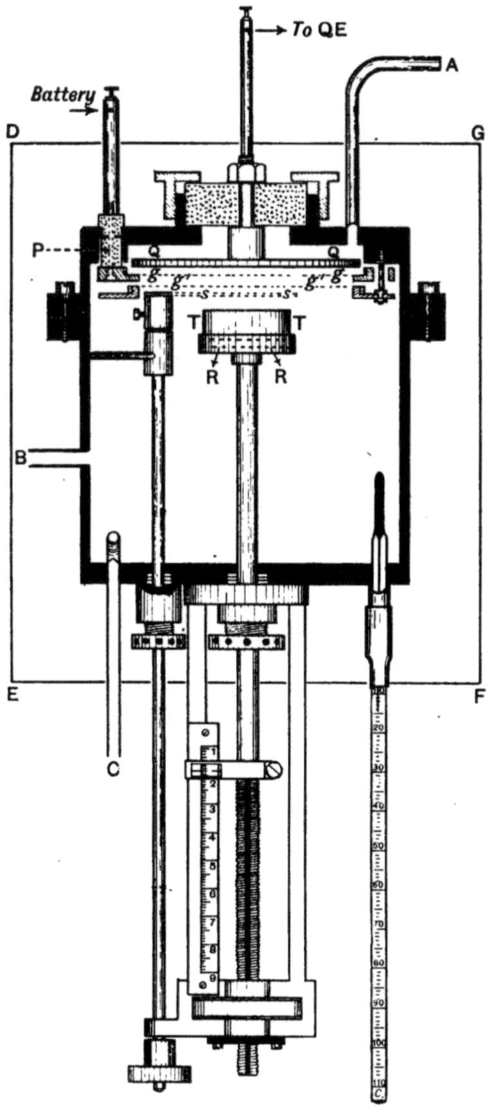

FIG. 6.—Apparatus for finding the range of α rays.

The plate *QQ* is connected to a Dolezalek electrometer, and the gauze *gg* to the battery, so that the gauze is at a high potential compared to the plate, which is initially earthed through the electrometer. The case of the instrument is earthed permanently, so that no electricity can pass over from the battery to the electrometer. In order to prevent accidental and troublesome electrostatic effects, the interior is so arranged that no lines of force end on *QQ* but such as have proceeded from metal surfaces of definite and unaltering potential. For this purpose the glass pillars which support *gg* are buried in cavities in the body of the case: if there should be any leakage over the pillars, there cannot be any resulting electrostatic action on the plate *QQ*.

An earthed gauze sheet *g′g′* is placed as far below the gauge *gg* as the plate *QQ* is above it. The electrical field is thus made symmetrical on the two sides of *gg*, and no ions from below *gg* can make their way into the chamber. It is to be remembered that there is heavy ionisation below the gauze, and none of it must contribute to the current measured. A screen *ss* of very thin copper is mounted on a rod which passes into the apparatus through a stuffing box and can be raised, lowered, and turned round. The screen can therefore be placed just over the radium tubes so as to cut off the α radiation, at whatever height the radium may be: and it can be turned aside so as to let the α rays pass. The difference between the currents with the screen on and off is the measure of the α ray effect, since the γ rays and most β rays can cross the screen. The case itself is made in two parts which are screwed together as shown, the line of division being placed so that the gauzes and the ionisation chamber all come off in one piece when it is necessary to open up the apparatus. The tubes *A*, *B*, and *C*, are for the purpose of filling the case with any desired gas or vapour. A thermometer is inserted, the precision with which ranges can be measured rendering it necessary to know the temperature within a degree. As it has to work upside down the top is knocked off, so that the air pressure holds the mercury together.

The outline *DEFG* represents an electrically heated oven. Sometimes it was necessary to work at a temperature as high as 70° C. in order to obtain the vapours of certain liquids at a sufficient density. Ebonite softened and yielded at this temperature and glass was used in place of it.

In some experiments in which all possible care was taken to work with pure gases, the case was surrounded by an outer case with an intervening space. The tubes, rods, and thermometer had to pass through a second set of stuffing boxes or air-tight joints in the outer case. The object of this arrangement was to nullify the bad effect of the small leaks through the stuffing boxes. No air could pass through and mix with the gas in the inner chamber if the pressure of the air between the outer and inner chambers was not allowed to exceed that of the gas.

CHAPTER III

THE IONISATION CURVE OF THE α RAY

WHEN the ionisation produced in the ionisation chamber of the apparatus just described is plotted against the distance between the radium and the chamber a curve is obtained of which Fig. 19 on p. 35 is an example. In this particular case the apparatus was filled with methane at a pressure of 64·3 cm. and a temperature of 18·5° C. Such curves show the four ranges of the α-particles from radium and yield information on various other points which we can now consider in detail.

Let A and B in Fig. 7 represent the upper and lower walls of the ionisation chamber and let C and D be the upper and lower surfaces of the radioactive layer. Assume that each α particle emitted by the radioactive material has the same initial velocity and the same range, and assume also, provisionally, that the particle moves in a straight line from start to finish and causes uniform ionisation along its track. Consider the action of such particles only as move normally to the layer, a limitation which is secured in the actual experiment. Let the distance between B and C be denoted by y, the depth of the ionisation

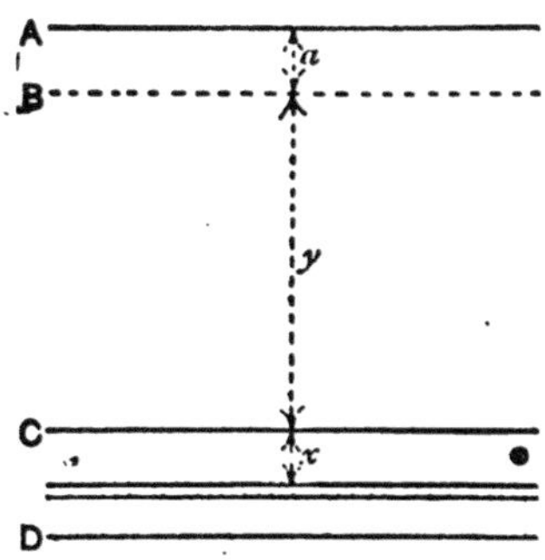

FIG. 7.

chamber by a (*i.e.*, the distance between A and B), and the range of the α particle in air by R.

Consider the α rays from a layer of width dx, at a depth x in the plate, and suppose that the material of the plate stops the α particle s times quicker than the air, so that the range in the material is only R/s. If R is greater than $y+sx$ the rays from dx will enter the ionisation chamber, but if R is less than $y+sx+a$ they will not reach the further side. In such a case the ionisation due to the α rays from dx is proportional to the length of that portion of their path which they complete within the chamber, that is to $R-y-sx$, and we may put it equal to

$$k(R-y-sx)dx.$$

If $R-y$ is less than a, so that the α rays from the very top of the radioactive layer cannot reach the top of the chamber, the whole ionisation is therefore

$$\int_0^{\frac{R-y}{s}} k(R-y-sx)dx=k.\frac{(R-y)^2}{2s} \quad . \quad . \quad . \quad . \quad . \quad . \quad . \quad . \quad \text{(i).}$$

If $R-y$ is greater than a, then the particles from all layers between $sx=0$ and $sx=R-y-a$ go right across the chamber, while from $sx=R-y-a$ to $sx=R-y$ they behave like those we have just considered. The ionisation is then

$$ka(R-y-a)/s + k\int_{\frac{R-y-a}{s}}^{\frac{R-y}{s}} (R-y-sx)dx=k.\frac{a\{2(R-y)-a\}}{2s} \quad . \quad . \quad \text{(ii).}$$

Supposing therefore that the radioactive layer is thick enough, that is to say thicker than a/s, the ionisation curve will consist of two portions at least, the first parabolic and represented by (i), the second a straight line represented by (ii). If we produce the straight line backwards to cut the vertical axis at T, as in Fig. 8, it follows from (ii) that $OT=y=R-a/2$, or $R=a/2+OT$. The range is therefore found by adding half the depth of the ionisation chamber to the intercept OT. In practice it is convenient to draw the scale on the instrument in such a way that it gives the distance from the top of the radium layer to the middle of the chamber, and the intercept OT as determined by the scale and the ionisative curve then measures the range directly.

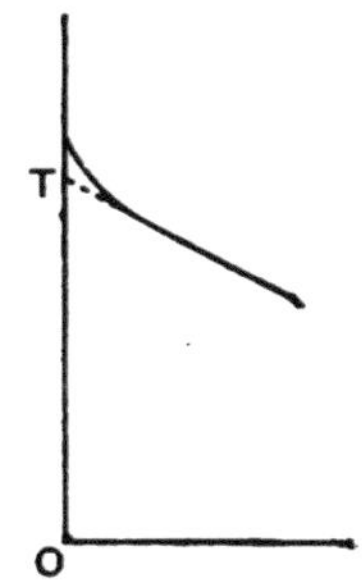

Fig. 8.

If there are n sets of α particles issuing from the radioactive material, and if it is assumed that these sets differ in speed and range only (not in number or in ionising power), the current through the chamber is given by

$$k\alpha(R_1 - y - a/2)/s + k\alpha(R_2 - y - a/2)/s + \ldots\ldots$$

where $R_1 R_2 R_3$. . are the ranges of the various sets, it being supposed that the chamber is within range of all the sets. This gives a straight line the inclination of which to the axis of x is proportional to n. Radium in radioactive equilibrium contains four substances emitting α particles, neglecting those products which grow so slowly that their amount in newly prepared radium is negligible. These substances are (1) radium itself, (2) radium emanation which is half transformed in 3·85 days, (3) radium A which has a period of 3·0 minutes and (4) radium C which has a period of 19·5 minutes. The last appears to be a complex substance, undergoing more than one transformation, but for our purpose this is immaterial. Radium B has a period of 26·7 minutes but sends out no α particle : radium D has a period of many years and we therefore neglect it and its successors. Since there are four substances emitting α rays in equal numbers, the ionisation curve should show four straight lines forming the sides of an equiangular polygon, the corners being rounded off. The curves shown in Figs. 3 and 4 show the sides and corners which we are led to expect : but they are scarcely worthy of a quantitative interpretation, because they were obtained with a crude form of apparatus and the use of thick radioactive films was abandoned before better apparatus was constructed.

The curve for a thick radioactive layer consists, as already shown, of a short parabolic portion and a straight line which runs down to meet the axis of x. In the case of a thin film the rays emerging from the lowest layer have still some part of their range to complete when they emerge, and the curve for a single set of rays must take the form of ABL in Fig. 9, the same conditions being assumed as before. The portion BL is upright because for all points upon it the rays all cross the chamber and the ionisation does not then depend upon the distance. Radium with its four sets of particles should give a curve consisting of four such units as shown in the figure, which will be clearly distinguishable from each other if the ranges are not too nearly equal.

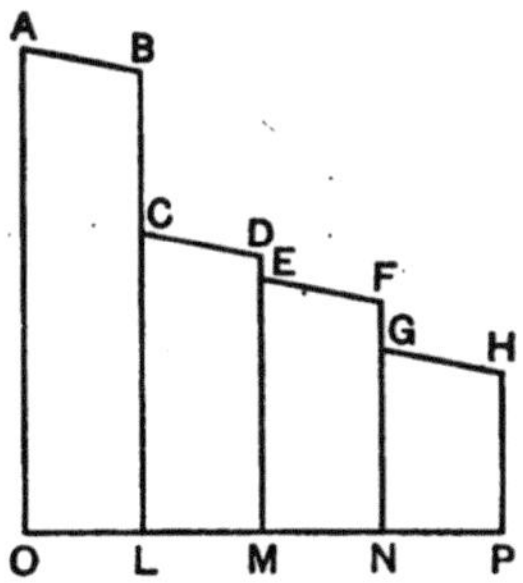

Fig. 9.

The actual curve in Fig. 10 shows a close correspondence with the anticipated form. The main difference is that the portions corresponding to BC, DE, FG, HP, are not upright but bend away from the axis of y for increasing values of the distance between the radium and the chamber. The inference is that it has been wrong to suppose the ionisation to be uniform along the track; it must increase as the α particle slows down. We can consider this point later.

From the curve we can find at once the ranges of the four sets of α particles. The greatest is 7·06 cm., this value being corrected for variations from the normal pressure and temperature of the air, viz., 760 mm. and 20° C. Later observations have shown this to be about 1% too small;

and have given the value 7·14 cm. The other ranges can be found in the following way, which illustrates at the same time an important feature of Rutherford's theory. In the figure the curve marked *E* was obtained when the radium was twenty-eight days old, by which time, according to the theory, the various substances in the radioactive layer should be emitting α particles at equal

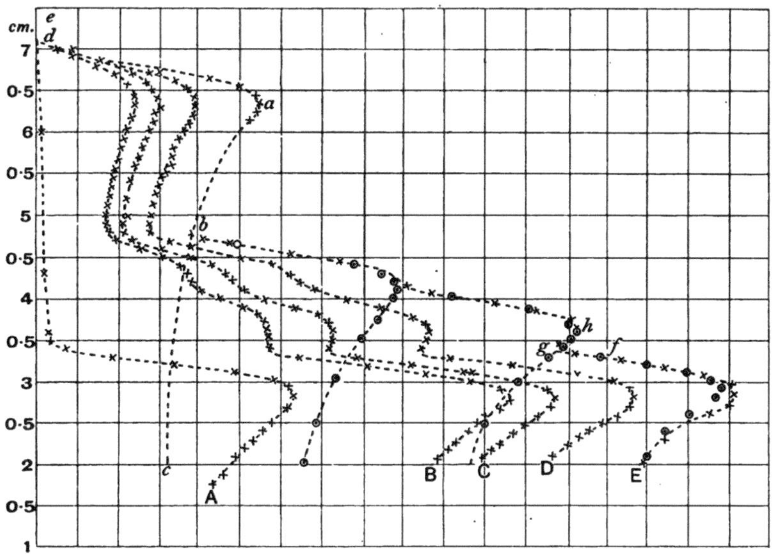

FIG. 10.

A.—Ionisation curve of Ra immediately after preparation.
B.— ,, ,, ,, after 64 hours.
C.— ,, ,, ,, ,, 90 ,,
D.— ,, ,, ,, ,, 140 ,,
E.— ,, ,, ,, ,, 28 days.

Experimental readings marked by crosses; calculated, by dots in circles. On the scale used the curve from *d* to *e* practically coincides with the bounding line of the figure.

rates. Let the curved side *ab* belonging to RaC be produced to *c* as shown. The line *de* represents ionisation due to a certain quantity of β and γ radiation which is intercepted by the aluminium screen (see Fig. 6), because like all the other abscissæ in the figure it is the difference between the readings when the screen is off and when it is on. When the radium is within range of the chamber this small quantity is included in the current which is observed. Its value must be nearly constant over all ranges, because it represents only a fraction of the full β and γ current, and the latter can be shown by experiment to vary little with the range, in this apparatus.

We may therefore consider that the figure bounded by the vertical line through *de* and the curved line *dabc* gives us the ionisation by the particles from RaC over the whole of their course except the two centimetres next the radium. This curve is now added to itself, being first lowered through 2·23 cm.; and the points represented by dots in circles represent some of the results of this addition. It will be seen how nearly these points lie on the actual ionisation curve, on which the experimental readings are marked by crosses. Again the curve *dabc* is lowered, this time by 6 mm. and added on to the sum of curves already obtained, and the new points which show the result of the addition are also marked as dots in circles. These also lie very nearly on the ionisation curve. For the last time the curve is lowered, by 7·3 mm. and added on; and again the calculated points lie on the experimental curve except just at the peak. Thus the full curve is formed by the superposition of four simple curves, alike in all respects but that of height, and the inference is that the four groups of α particles are alike in all respects but that of initial velocity. The range of the α particle of least range is 3·50 cm.; for the curve *fg* produced meets the curve *hg* also produced at a height of 3·40 cm., in this figure the distance from radium to gauze, and one millimetre is to be added because the chamber was two millimetres deep. The ranges of the other α particles must therefore be 4·23, 4·83, and 7·06.

The differences in height of the four portions of the curve show that the initial velocities of the four sets of α

particles are all different. The equalities in width show that there are equal numbers of α particles in the four streams and this is a most important feature of Rutherford's theory.

In order to find out which portions of the curve represent the effects of the particles from the different radioactive substances we can make use of the differences in physical properties of the substances. The radium layer is raised to a red heat and its ionisation curve is determined with as little loss of time as possible. It is found that the two middle steps have disappeared and the highest step soon follows suit. This is shown in Fig. 11, which represents the results of the original experiment. The apparatus used for the experiment was of an early pattern and the results are somewhat rough, but they are quite clear. It was never worth while to repeat the experiment because the conclusion drawn from it was afterwards fully confirmed by Rutherford in another way. The curve *I* was obtained about thirty hours before the heating and is not quite as far to the right as it would have been if it had represented the true state of things just before the experiment began. The radium was raised to a red heat, and the curve marked *II* represents measurements then taken with decreasing distances between the radium and the chamber. The distance was gradually increased again and the curve marked *III* was obtained, the whole of the readings taking a little more than half an hour. The straight line represents effects due to β rays and other

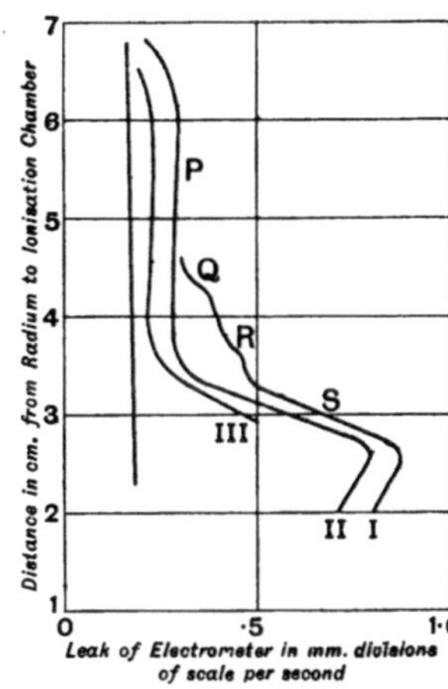

FIG. 11.

causes, and α ray effects are to be measured to the right of it.

Now according to Rutherford's investigations the emanation should be quickly removed by a red heat: and RaA should practically go with it since it is derived directly from the emanation and has only a life of two or three minutes. It all disappears in the few minutes that are occupied in replacing the de-emanated layer in the apparatus and once more commencing measurements. The two middle steps must therefore refer to the emanation and to RaA, and the highest step to RaC. This experiment did not decide as to the allotment of the ranges to emanation and RaA; but subsequently an observation by H. W. Schmidt (*Phys. Zeit.*, 1905, p. 897) showed that the range of RaA was nearly two-thirds of that of RaC. Consequently the 4·83 mm. must be given to RaA and the 4·23 mm. to the emanation. In the series of derivatives, therefore, each of the four radioactive substances emits an α particle with more violence than those that precede.

When the plate of radium has thus been heated and its radioactive products driven off, it is to be expected that the radioactive equilibrium will be restored again in due time, the rate of recovery being already known from Rutherford's work. The succession of figures *A*, *B*, *C*, *D*, and *E* in Fig. 10 shows the realisation of this expectation. The rate of growth of the emanation is slow: it arrives at half its final value in 3·85 days as already said: whereas RaA, RaB, and RaC have such short lives that their rate of emission lags very little behind that of the emanation. Consequently the three steps in the figure due to emanation, RaA, and RaC are very nearly of equal width throughout the whole time of their growth.

The method of finding the ranges of the α rays of radium which has been described in the preceding pages has also been employed to find the ranges of the α particles of other substances. Hahn found in this way the ranges of the thorium derivatives, using the substance radio thorium which had just been separated. (*Phil. Mag.*, July, 1906.) He found also the ranges of the actinium rays. (*Phil. Mag.*, Sept., 1906). A very interesting feature of the thorium experiments was the discovery that the substance known as ThB had a double transformation. The curve for the freshly prepared substance had the form given in Fig. 12 and this shows two steps unmistakably. The period of one must be so short that it is impossible to obtain it by chemical means: if it is separated it disappears immediately.

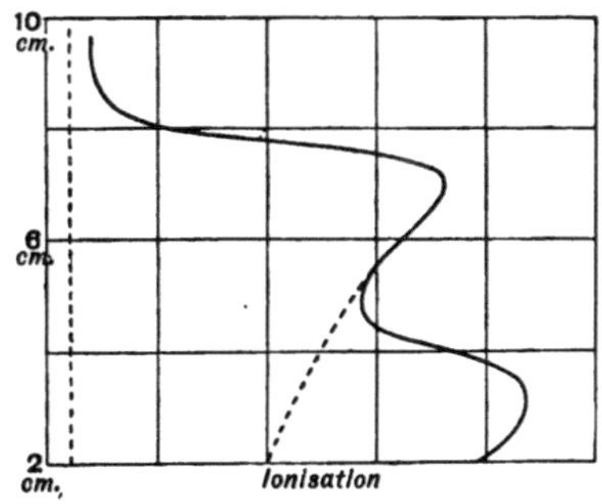

FIG. 12.

Levin found the range of the α particle of polonium to be 3·86 cm. (*Amer. Journal of Science*, July, 1906).

Uranium and also thorium in its ordinary condition cannot be treated by this method since the α-radiation is too weak to allow the use of the set of tubes which is an essential part of the apparatus. There is plenty of ionisation by the α rays above a thick layer so long as the tubes are not employed, and it is easy to investigate experimentally the result of placing sheets of absorbing material upon the radiating layer. The results to be expected may also be calculated in terms of the range of the α particles and the form of the regular ionisation curve, provided there is only one set of α particles in action. There is then no difficulty in comparing the calculated with the experimental results and so finding the range. The formulæ are rather long and complicated, and as they are given in full in the *Philosophical Magazine* for June, 1906, it will be sufficient to reproduce here the diagrams which represent them graphically. (Fig 13.)

The curve *A* refers to the case where the radioactive material is so thick that the rays from the bottom of it do not emerge. The abscissæ represent the thickness of the absorbing layer measured by the proportionate amount of its range which an a particle would lose in going through

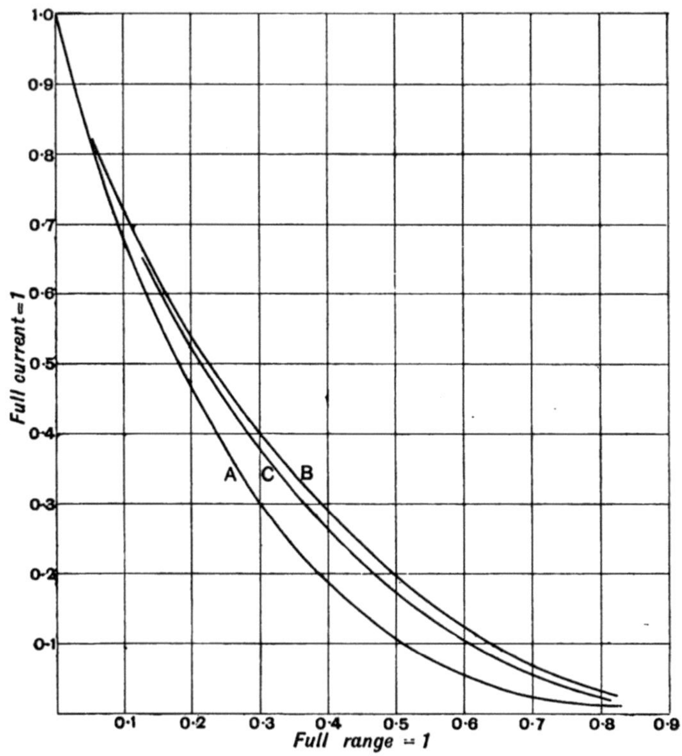

Fig. 13.

it normally. The ordinates represent the ratio of the current for a given absorbing screen to the current without any screen. In the case of *B* the radiating layer is very thin, while *C* is an intermediate case when the thickness of the layer is equivalent to half the range. As far as possible the calculations were made to take account of the increase of ionising power towards the end of the range, and this introduced a little uncertainty (*loc. cit.* p. 757), for it was necessary to know the law of the increase before this could be done correctly and the law was not well known at that time. But the corrections are only small, and the curves are therefore still quite good enough to be of use. The range cannot of course be found by this method with an accuracy even approaching that of the former. By its means I found the ranges of the α particles of radium, thorium (actually radiothorium), and uranium to be approximately equal: not a good result as it now appears. For though the range of thorium, 3·86 cm. (Hahn), is not far from that of radium, 3·50, yet the range of uranium is much smaller according to Rutherford and Geiger, who give the value 2·7 cm. (*Phil. Mag.*, Oct., 1910, p. 698). I think that the error must have been due to impurity in the material which I used.

These curves are interesting from an entirely different point of view. In the earlier experiments in which the absorption of the α radiation by various materials was investigated, thin sheets of aluminium, gold, tin, or other material were laid upon the radioactive layer and the current then measured. The curve *A* shows what the result would be in such a case, and it is strange how closely the curve seems to follow an exponential law. When the thickness of the absorbing layer is one-, two-, or three-tenths of the full range of the particle in the material of the layer, the ionisation current is 0·675, 0·46, and 0·30 respectively of the original value. These members lie very near to the geometrical progression 1, 0·675, 0·455, 0·305: and it is no matter of surprise that the results were supposed to establish an exponential law. It seemed natural to adopt such a law in respect to radiation, and in consequence the true nature of the absorption of α rays was overlooked. As a matter of fact, the exponential law of absorption has an immediate meaning only in the case when each absorbing layer deducts a definite fraction of the energy of the stream of radiation, leaving the remainder unaltered in every respect but that of quantity: and this not merely of the average quantity but of the quantity moving in each direction, supposing the stream to be other than unidirectional. Even in the case of monochromatic light, exponential absorption only takes place when the light consists of a parallel beam passing through a plane absorbing screen of increasing thickness.

Geiger has recently used a method of determining ranges which is specially fitted for use with substances whose activity is small (*Phil. Mag.*, Oct., 1911). He places the material under investigation at the centre of a silvered glass bulb and measures the ionisation of the air which the bulb contains. The pressure of the air can be varied, and so long as the *α* particles do not reach the walls the ionisation is practically constant. As the pressure is lowered, a point is reached at which the *α* particles reach the walls, and from that point onwards the ionisation steadily declines, being proportional to the pressure. This point is easily determined and the range in air at some standard temperature and pressure can be found. Geiger gives the following values for the ranges of certain substances, reduced to 76 cm. and 0° C. I have placed alongside the corresponding values for 76 cm. and 20° C. in order that comparison may be easily made with the values already given in this chapter.

TABLE I.—*Ranges of Certain α Particles.* Geiger.

	76 cm. 0° C.	76 cm. 20° C.		76 cm. 0° C.	76 cm. 20° C.
Uranium . . .	2·58 cm.	2·75 cm.	Polonium . . .	3·58 cm.	3·84 cm.
Ionium	2·84 ,,	3·05 ,,	Thorium . . .	2·58 ,,	2·75 ,,
Radium . . .	3·13 ,,	3·36 ,,	Radiothorium .	3·67 ,,	3·94 ,,

In the same paper Geiger confirms and extends a very remarkable observation of Rutherford's, that the range of the *α* particle is, in general, greatest when the life of its parent substance is shortest.

CHAPTER IV

INTERPRETATION OF CERTAIN PECULIARITIES OF THE α RAY CURVE

WE may now consider in greater detail the interpretation of the peculiarities in the form of the ionisation curve, which we can divide for this purpose into the upper slope *AB*, and the more vertical portion *BC*.

The existence of the slope *AB* is sometimes to be ascribed in part to the thickness of the radioactive layer, as has been already explained. When the layer is so thick that some α rays lose, say, a centimetre of their subsequent path in the air through loss of energy in crossing the layer, then the corner *B* is correspondingly depressed and rounded off. But it is found that when the layer is made very thin indeed, as in the case where a deposit of RaC is used as the source of rays, there is still a slope at the top of the curve and the point *B* is about half a centimetre lower than *A* in air at normal pressure and temperature. Such a deposit is so thin that the first explanation is quite insufficient. McClung was the first to measure in this way the range of RaC and to draw its ionisation curve (*Phil. Mag.*, June, 1906).

FIG. 14.

More recently Geiger has repeated the experiment very carefully by a method which is described in the Proceedings of the Royal Society, 82, p. 489 (May, 1909). Its essential difference from the original method consisted in the use of a gas at very low pressure in the ionisation chamber, which was cut off from the air between the radium and the chamber. This was equivalent to the use of a very shallow ionisation chamber filled with ordinary air and in consequence the ionisation due to the *a* rays was more exactly studied at different points of the *a* ray path. The curve thus determined is reproduced in Fig. 15; and is doubtless the most exact curve yet obtained. The sharpness of the peak at *B* may be assigned mainly to the narrowness of the ionisation chamber: and perhaps in part also to the ease with which all the ions made in air at low pressure, or in hydrogen which was also used, can be gathered up by the applied potential. (See later, p. 73.)

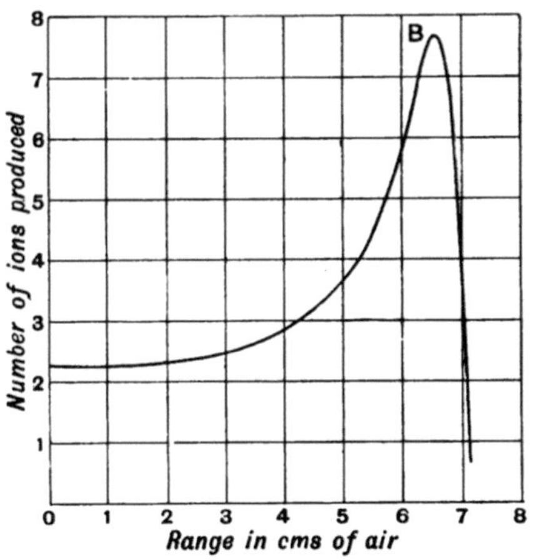

Fig. 15.—Ionisation curve (Geiger.)

We have therefore to explain this slope as due to some property of the *a* rays and not to experimental arrangements. It might be supposed that it is due to want of uniformity in the velocity with which the *a* particles are expelled from their source. From a certain experiment of Rutherford's, however, it is clear that this cannot be done, for the photographic image of a pencil of *a* rays is as sharp and narrow when deflected by a magnetic field as when no field has been applied; and Geiger has repeated and confirmed this observation with the especial object of testing the possibility of such an explanation. (*Proc. Roy. Soc.*, 83, p. 512.)

The *a* particles are therefore all projected with the same initial velocity: if they complete their paths at various distances from their source they must have had different experiences on the way. Since each one produces ionisation by a very large number of independent acts accompanying its passage through a very large number of molecules, the effect cannot be ascribed to differences in the amounts of energy spent in ionisation.

It is possible that a sufficient explanation can be found in the "scattering" of the *a* particles which Geiger has investigated (*Proc. Roy. Soc.*, 83, p. 492). The particles can be turned aside from their rectilinear path on extraordinary occasions during their passage through matter. The amount of this deviation is so small in the case of the passage through air, or through thin sheets of metal when the speed is high, that the original conception of the movement of the *a* particle is hardly affected at all; and the corresponding experiments which I founded on the conception do not directly show the influence of scattering. Rutherford and Geiger devised another method of following the path of the *a* particle in which the effect became plain: they were able to observe the action of a single particle, whereas my method dealt with the effect of a stream of *a* rays (Rutherford and Geiger, *Proc. Roy. Soc.* 81, p. 141; Geiger, *Proc. Roy. Soc.*, 81, p. 174; and *Proc. Roy. Soc.*, 82, p. 486; Geiger and Marsden, *Proc. Roy. Soc.*, 82, p. 495; Geiger, *Proc. Roy. Soc.*, 83, p. 492, and p. 505).

The method consisted in the observation of the distribution of the *a* particles at any point in the path of a pencil originally narrow by means of the scintillations produced by the impacts upon a phosphorescent screen. A sketch of the apparatus used in some of the experiments is given in Fig. 16, which is taken from Geiger's paper (*Proc. Roy. Soc.* 83, p. 493). It is found that each particle makes one, and only one, scintillation when it strikes the screen. Observation of the number and position of the scintillations is a perfectly trustworthy index of the distribution and number of the particles. It is only at the very end of their range, when they have but two or three millimetres left to go, that the scintillations cease to be visible. When they have no screen to pass through, and the apparatus is exhausted of air, the *a* particles from an active deposit at *A* always fall within the geometrical shadow of the hole at *D*; but as soon as a screen is placed at *E* or *D*, the phosphorescent spot

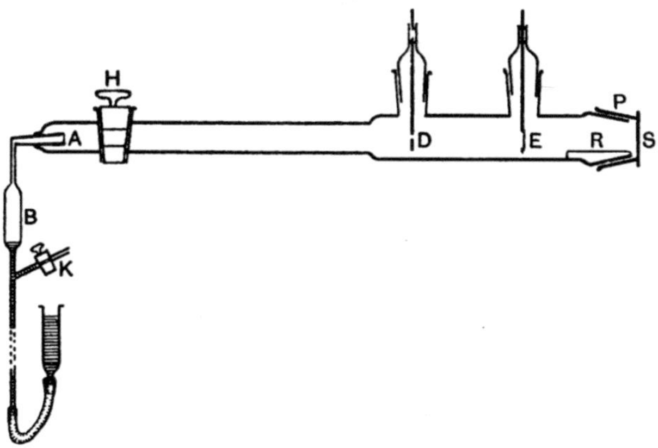

FIG. 16.—*Ra* emanation is compressed in the small conical tube *A*, until a sufficiently strong radioactive deposit has been made on its walls. It is then withdrawn. A small pencil of *a* rays passes down the tube through the openings provided and falls upon the phosphorescent screen at *S*. The separate sparkles made by the individual *a* particles are observed through a microscope.

at *S* opens out, showing that the particles have been scattered to a certain extent in passing through the screen. The amount of the scattering may be expressed in terms of the probable angle of scattering (K) in going through a given screen, and curves can be drawn, as in Fig. 17 (*loc. cit.*, p. 497) in which the ordinate of a point on any one of the curves represents the probable number (relatively) that will be scattered through the angle given by the abscissa. One hundred gold foils reduce the range in air to nearly half its original value, and the figure shows that after passing through such an amount of gold, they are most probably deflected about 7°. From these experiments it is clear that the scattering becomes greater as the particles slow down, and Geiger concluded that the probable scattering varied approximately as the inverse cube of the velocity. This being so, the scattering must become very marked towards the end of the range. It may be added that Geiger and Marsden (*Proc. Roy. Soc.*, 82, p. 495) showed that when a stream of α particles fell on a gold plate, about one in eight thousand was so far deflected as to emerge from the incidence side of the plate.

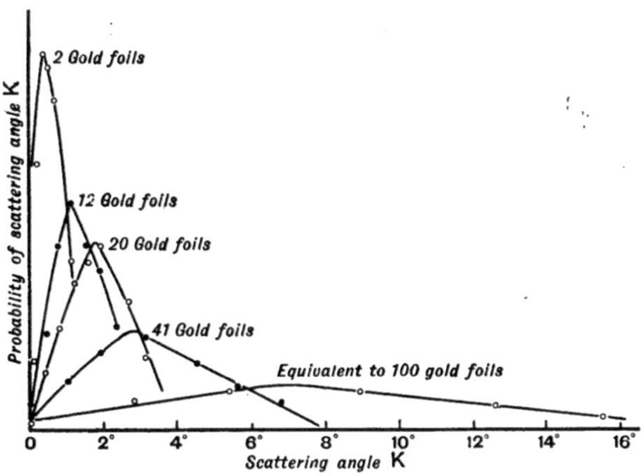

FIG. 17.—Scattering curves (Geiger).

The results of the investigation are summed up as follows:

(1) The most probable angle of scattering increases for small thicknesses in approximate proportion to the square root of the thickness of matter traversed by the α particle. For greater thickness the scattering increases more rapidly.

(2) The probable angle through which the α particle is turned in passing through an atom is proportional to its atomic weight. The actual value of this angle in the case of the gold atom is about $\frac{1}{200}$ of a degree.

(3) The most probable angle of scattering increases rapidly with decreasing velocity of the *a* particle, being, to a first approximation, inversely proportional to the third power of the velocity.

All these results are direct deductions from experiment, except perhaps the second part of (2) in which it is assumed that the law enunciated in the first part of (1) holds down to extremely small thicknesses of gold, although the thinnest foil obtainable puts about 160 atoms in the way of an *a* particle trying to cross it.

The paths of *a* particles through the air must be something like those which are shown (with exaggerated deflections) in Fig. 18: towards the end they are knocked about in passing through the atoms they meet: at the very end they are merely helium atoms jostled in a crowd of other atoms. Perhaps this is sufficient to account for the inequalities in the ranges of the *a* particles of the same stream, and it must be a contributory cause at least.

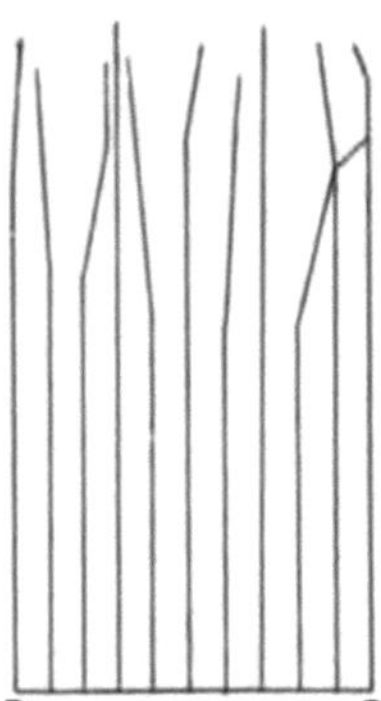

R R

FIG. 18. — Suggested forms of the paths of *a* particles projected upwards from radium at *R R*.

On this hypothesis the proportion of the sloping portion at the top of the curve to the rest of the curve ought to vary considerably with the atomic weight of the gas passed through. This expectation is partly realised at any rate. Methane (CH_4) contains a number of small atoms as compared with air, and the hydrogen in it is responsible for half its absorbing effect (see later). Therefore, on this view, the slope should not take up quite so much of the curve. The ionisation curves of methane and air are drawn side by side in Fig. 19, p. 35, and the expected effect is clearly shown. The inclination of the top part of the curve (for each set of *a* rays) is less in methane than in air. This is also to be deduced from a comparison of the ionisation curves of air and hydrogen made by Taylor and shown on p. 368 of Vol. 28 of the *American Journal of Science*, Oct., 1909. It is reproduced on p. 52 below.

There is another possible explanation. It may be that the ionising effects of a single *a* particle reach a maximum at a certain speed corresponding to a point on the path represented by *B*, Fig. 14, after which they decline. Wilson's photographs of the "fog tracks" of *a* particles (Fig. 53) certainly support this idea.

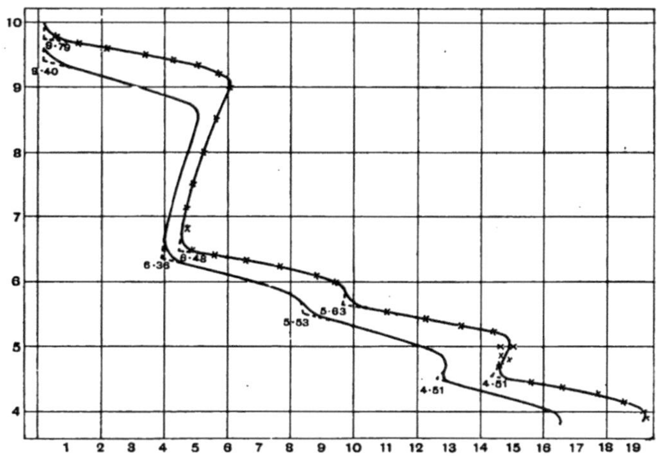

FIG. 19.—Upper curve; methane at 18·5° C. and 64·3 cm. Lower curve; Air at 15·5° C. and 57·0 cm. (*Phil. Mag.*, Sept., 1907).

We may next consider the form of the main part of the ionisation curve of a pencil of initially homogeneous *a* rays—*i.e.*, the portion *BC* in Fig. 14. The remarkable slope back of the curve from 6·5 cm. to the zero point on the range—to take the case of RaC—shows clearly that the ionisation produced by the particle increases as the speed diminishes throughout this range. The exact relations between the velocity, the range yet to be completed, the rate at which ionisation is being produced, and the rate at which energy is being spent are all subjects of interest, because they alone can help us to understand what happens during the passage of the *a* particle through the atom.

The form of the curve gives the relation between the rate at which ionisation is being produced, and the range yet to be completed: experiments on the magnetic deflection of the α rays show the relation between the latter quantity and the velocity.

As the result of his original experiments on the magnetic deflection, Rutherford showed that the velocity was proportional to $\sqrt{(r+1\cdot 25)}$, where r was the uncompleted range. From the form of the ionisation curve it appeared (Bragg, *Phil. Mag.*, Nov., 1905) that the ionisation at any point in the path was proportional to $1/\sqrt{(r+1\cdot 33)}$. The two numbers in these results may be taken to be the same within the errors of experiment and from a combination of them it follows that the ionisation is inversely proportional to the velocity. Both these primary formulæ are sufficiently accurate to justify the deduction, which has indeed been confirmed by later work. Neither of them applies, however, to the extreme end of the range, say, the last 5 or 10 mm., where there are the most important and rapid variations, both in the velocity, and in the rate at which ionisation is produced. This is now clear, principally through the results of the investigation of the scattering effect which has been described above. It is just in this part of the curve that scattering becomes prominent, and the neglect of it leads to error. In the earlier experiments the full effect of the scattering was not realised.

Rutherford found, for instance (*Phil. Mag.*, Oct., 1906) that when the α particles had passed through 14 aluminium foils, equivalent to nearly 7 cm. of air, and had all but completed their range, they were still deflected by a magnetic field through so small an angle that they seemed to possess nearly half their initial velocity. But they were unable to pass through another sheet of foil. He therefore concluded that the α particle completed its range with much of its energy still unspent, and this is implied in the form of expression, $\sqrt{(r+d)}$. Various suggestions were made at the time in order to explain this surprising result. It was supposed that the α particle then picked up enough negative electricity to become neutralised, and that when neutral it could not ionise, or, again, that it was absorbed in some atom while still going at this great speed—a suggestion of the greatest interest since it suggested a building-up of the atom, which must exist but has not yet been found. In some way an explanation had to be given for the very strange fact that exactly at a certain speed the α particle lost the power of ionising any kind of atom or molecule, of exciting phosphorescence, and of acting on a photographic plate.

The scattering of the α particles which Rutherford was the first to observe (*Phil. Mag.*, July, 1905) suggested possible sources of error in the magnetic deflection experiments, and finally Geiger undertook to revise them very carefully. It then appeared that the formula $v = k\sqrt{(r+d)}$ should be replaced by $v = k'r^{1/3}$, which gives nearly the same form of curve over a considerable range of values of r, so long as the latter are not too small and not too large. It gives, however, a zero velocity at the end of the range, and makes a fundamental difference to the theory. The experimental distinction between the two depended on observations at the very end of the range where the scattering is so important, and the scintillation method is then the most effective that can be employed. The fifteenth sheet of foil in Rutherford's experiment had so scattered the α particles that their career as a stream of particles was ended, and even if some got through they were not in a condition to produce an obvious impression on the photographic plate. They seemed to have disappeared while yet their energy was partly unspent. In these circumstances the same change of formula can equally well be made in the case of the ionisation curve, and we still conclude that the ionisation varies inversely as the velocity. Geiger puts $v^3 = ar$,

$I^3 = a'/r$, where v is the velocity, I the ionisation, r the range to be completed, and a and a' are constants. The rate at which energy is spent is proportional to vdv/dr and therefore to $r^{-1/3}$ and to the ionisation; that is to say, the ionisation produced is proportional to the energy spent.

Since Iv is a constant, it appears that the ionisation is proportional to the time which the α particle spends within the atom. We are here comparing the effect of varying the speed while the nature of the molecule is kept the same, but, curiously enough, as we shall see later, if the speed considered is kept the same and the nature of the molecule varied, there is still a rough proportionality between the ionisation and the volume of the molecule, and therefore the time taken by the α particle to pass through it. These facts seem to suggest that the ionisation is a consequence of the presence of the α particle within the atom and depends in its amount on the time for which that presence lasts.

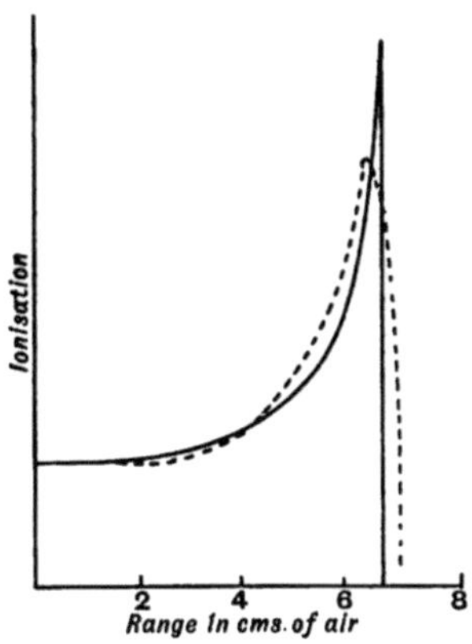

FIG. 20.

Fig. 20 is taken from Geiger's paper (*loc. cit.*, p. 513). It shows in firm line the theoretical form of the ionisation curve as given by the formula $I_3 = a'/r$, unaffected by scattering, the width of the chamber, and the geometrical arrangements generally. The dotted curve shows the results of the experiment on the form of the curve already referred to. The agreement is certainly very satisfactory.

CHAPTER V

STOPPING POWER

WE have seen that the path of an α particle through a gas is very nearly a straight line, except at the extreme end of its range. The length of the range is found to depend not only upon the density of the material penetrated, but also on its nature, and it becomes a matter of great interest to discover the relations between the range and the nature of the substance, since in this way some light may be thrown on atomic structure.

In the case of such substances as can be obtained in the form of very thin sheets, it is convenient to place them over the radioactive material in the apparatus of Fig. 6. The particles cross the sheet normally, losing range in doing so. The ionisation curve being drawn, it is found that it has dropped throughout the whole of its length, as shown in Fig. 21 (Bragg and Kleeman, *Phil. Mag.*, Sept., 1905). The curve marked A is a normal ionisation curve in air: that marked B represents the result of placing a sheet of silver weighing 0·0021 gr. per $cm.^2$ in the path of the α rays. The two curves exactly resemble each other in all but height. The silver has partly stopped the α particles, cutting off so much range from each one: it is equivalent in "stopping power" to a little less than a centimetre of air. The effect of interposing a thicker sheet of silver weighing 0·0097 gr. per cm.^{2} is also shown in the same figure.

This effect is really an example of the main principle on which we have been proceeding, but it is interesting to observe the behaviour of a solid absorbing screen as well as of a gaseous one, and to find such a satisfactory illustration

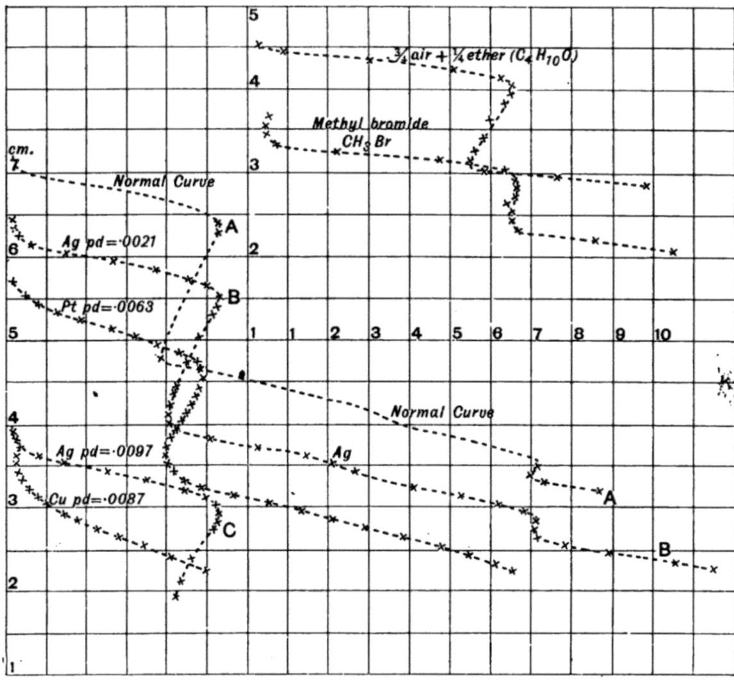

FIG. 21.—Ionisation curves showing effects of placing certain metallic screen in the path of the *a* particle; also of substituting certain gases for air.

of independence of physical state. Let us consider for a moment the meaning of such a result as this.

If any of the *a* particles had been held back by the metal sheet the curve would have contracted sideways; if any had lost more range than others, or if scattering had had an appreciable influence, the slope of the top of the curve would have been altered. Since nothing is to be seen of either of these effects, the simple theory already given must be sufficient. The *a* particle must pass through all the atoms it meets, losing energy in doing so. It cannot pursue a zig-zag course between the atoms of the silver because, having no intelligence, it cannot recover a line once lost. Nor can it push the atoms it encounters out of its way, for they are heavier than it is itself, and it meets with hundreds of thousands. It must go through the atoms it meets, and in this mutual penetration of the atoms we have an exceedingly instructive revelation of their forms. In the ordinary movements of the molecules of a gas two atoms may collide while their centres are so far from each other that the "free path" is extremely short. The range in such a case, for we may well so term the free path, is about 10^{-5} cm. But now an exceedingly close approach of the centres is possible, since an *a* particle has a range of several centimetres. No doubt the immense speed of the *a* particle is responsible for the change. The impenetrability of one atom by another is only a fact when the mutual velocity is small, say, 10^5 cm. per sec.; it ceases to exist when that velocity is sufficiently increased, say, ten thousand times. We shall find similar suggestions that the impenetrable condition is only an accidental one throughout the whole range of the phenomena of the new radiations.

The experiment shows that it is possible to express the stopping power of a silver sheet as equivalent to that of a stratum of air of thickness found by experiment. A number of such measurements were made by Kleeman and myself in 1905, and our results are given in the following table (*Phil. Mag.*, September, 1905, p. 332). The second column gives the product of the thickness of the metal film and its density, the third the corresponding drop in the curve multiplied by the density of air, and the fourth the ratio of these two products.

TABLE II.—*Stopping Powers of Metal Sheets.*

I.	II.	III.	IV.	V.	V./IV.
Gold	0·0121	0·00396	3·05	14·2	4·65
Platinum	0·00633	0·00192	3·29	14·0	4·25
Tin	0·0051	0·00212	2·41	10·85	4·50
Silver	0·00967	0·00402	2·41	10·4	4·30
Copper	0·00873	0·00492	1·78	7·96	4·45
Aluminium	0·00258	0·00209	1·23	5·15	4·20

It is remarkable that the numbers in the fourth column are nearly proportional to the square roots of the atomic weights of the metals. To bring this out more clearly the square roots are shown in a fifth column and in a sixth the ratios of the numbers in the two previous columns.

Moreover air itself falls approximately into line with the metals. Its ionisation ratio should be entered in the fourth column as unity and the average square root of its atomic weight as $(4\sqrt{14}+\sqrt{16})/5=3{\cdot}79$. The corresponding entry in the last column should be also 3·79.

Also hydrogen is not far away from the other substances. At the time of which I am writing Strutt had measured its ionisation by a rays and found it to be 0·226 of that of air. Assuming that this meant that a layer of hydrogen 1 cm. thick had the same effect on the a particles as a layer of air 0·226 cm. thick, then the ratio of the product of the thickness and the density in the case of the hydrogen to the similar product in the case of the air is $1/14{\cdot}4\times0{\cdot}226={\cdot}31$. The square root of the atomic weight being 1, we were able to put in the last column $1/0{\cdot}31=3{\cdot}2$ as the proper figure for the hydrogen. It was satisfactory to be able to include hydrogen, not only because of its great difference in density, physical condition, and atomic weight from most of the other substances, but also because it had hitherto seemed to be anomalous in its behaviour to the new radiations.

Since the atomic weight appeared to be the one quality of importance, it seemed better to express our results more directly in terms of it. We therefore compared the stopping power of the silver with a stratum of air, not of equal weight but containing equal numbers of atoms. For equal weights silver stops 1/2·41 times as much as air: but for equal numbers of atoms it stops 108/14·4 × 2·41 times as much. This ratio, 3·11, we called the stopping power of the silver atom, referred to the air atom as standard: the latter being taken to have an atomic weight of 14·4 and an atomic square root 3·79.

The dependence of the stopping power on the atomic weight alone suggested a point of theoretical importance. If the physical condition counted for so little that a gas (such as hydrogen) and a metal (such as gold) could be compared without taking account of the fact that the one was a heavy metal and the other a light gas, then probably chemical association was of as little account. The stopping power of an atom would not depend on its association with other atoms in the molecule; and the stopping power of a molecule would be simply proportional to the sum of the square roots of the weight of the atoms which it contained. We tested this conclusion by finding the stopping power of methyl bromide (CH_3Br), in which the heavy bromine atom introduced a great differentiation between the weight and the square root of the weight. If the "absorption" in passing through a gas were proportional to its density, as the evidence had previously been thought to show, the range of the particle from RaC would be only about 7/3·28 or 2·3 cm., since methyl bromide was 3·28 times as heavy as air. But if the square-root law were true, and the stopping power of the molecule were the sum of the stopping powers of its atoms, then the range would be calculated thus:

$$\text{Stopping power of air} = \text{const} \times \frac{8\sqrt{14} + 2\sqrt{16}}{5} = 7{\cdot}58.$$

$$\text{Stopping power of } CH_3Br = \text{const} \times (\sqrt{12} + 3\sqrt{1} + \sqrt{80}) = 15{\cdot}41.$$

Hence, range of RaC in $CH_3Br = 7{\cdot}06 \times 7{\cdot}58/15{\cdot}41$.
$= 3{\cdot}4$ cm.

The stopping power of the molecule $= 15{\cdot}41/7{\cdot}58$.
$= 2{\cdot}03$

The curve obtained is shown in Fig. 21, the range of the *a* particle of RaC being 3·32. This was the distance from the radium to the gauze of the ionisation chamber. Half the depth of the chamber was to be added, and the true range was therefore one millimetre more, or 3·42. Allowing for a small amount of air that had got through the stuffing boxes and was measured by a determination of the density of the gas after the experiment—it amounted to one part in 75 in molecular weight equivalent—we found the stopping power *s* from the equation

$$\frac{1+75s}{76} = \frac{7{\cdot}06}{3{\cdot}42} = 2{\cdot}07$$

$$\therefore s = 2{\cdot}09.$$

The close agreement with the calculated value was thus in good accord with our hypothesis.

Many experiments of this kind were performed and led to equally satisfactory results, so that it became justifiable to assume the additive law to be true, and to use it in finding the stopping powers of some atoms which could only be examined in molecular association with others, such as chlorine, bromine, iodine, and sulphur.

The following table contains the results of some measurements of the stopping powers made at this and subsequent times :—

TABLE III.—*Atomic Stopping Powers.*

	s.	$\sqrt{w}$.	$s/\sqrt{w} \times 10$		s.	$\sqrt{w}$.	$s/\sqrt{w} \times 10$
H	0·24	1·00	2·40	Cu	2·46	7·96	3·09
C	0·85	3·47	2·46	Br	2·60	8·93	2·91
N	0 94	3·47	2·51	Ag	3·28	10·37	3·16
O	1·05	4·00	2·62	Sn	3·56	10·9	3·26
Al	1·495	5·20	2 87	I	3·44	11·2	3·07
S	1·76	5·65	3·12	Pt	4·14	13·95	2 97
Cl	1·78	5·96	2·99	Au	4·22	14·0	3·01
Fe	2·29	7·48	3·07	Pb	4·27	14·35	2·98
Ni	2·44	7·65	3·19				

The second column gives the stopping powers: the third the square root of the atomic weight. The additive law which holds for these constants implies that the opposition which any atom offers to the movement of the α particle is not modified by the presence of neighbours, even of the same molecule. We may even go further and say that the particle only acts on one atom at a time: since the scattering of the rays has not so far been found to obey any other rule. We shall find, indeed, that throughout the series of actions between the new rays and matter the collision between the ray and the atom is an extremely local event, in which it is impossible for several atoms to combine together and so have the effect of a larger one. Chemical association counts for nothing. Curiously enough, we shall see later that the ionisation produced in a gas by a given expenditure of α ray energy *does* depend to some extent upon the form of the molecule; and this would imply that the additive rule may apply strictly to the expenditure of energy in the molecule, but not so strictly to the result of that expenditure. So far, however, as the action of matter upon the rays is concerned, physical and chemical conditions are quite without influence; and to any of these energy carriers matter is merely a collection of atoms of which the geometrical arrangement is of secondary importance. There is, for example, no such phenomenon as reflection or refraction—there is only scattering.

A special series of experiments were undertaken by Dr. W. T. Cooke and myself in 1908 in order to make as careful a test of the additive law as we could accomplish. It was for this purpose that we surrounded the inner chamber of Fig. 6 with the outer casing described on p. 16, wishing to keep our gases free from any admixture with air. Great pains were taken to make gases initially pure, in which task we were most materially helped by the gift of a liquid air machine from the same generous source as before—that from which I had received the radium.

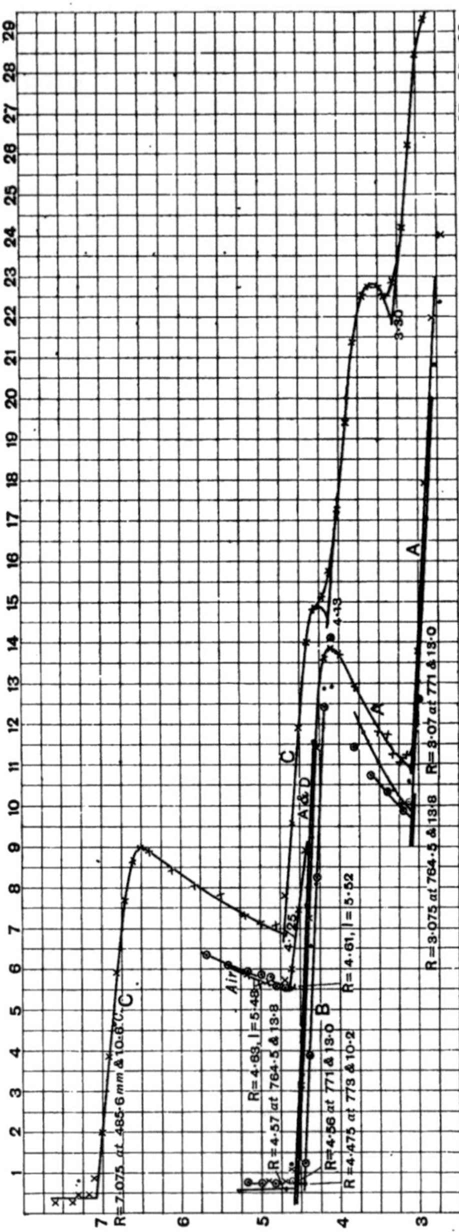

FIG. 22.—Ionisation curves in ethane (C_2H_6), used in finding the ethane figures in Tables IV and V. Pressures in mm. of Hg; temperatures in degrees Centigrade.

Ranges could easily be read to a fifth of a millimetre, and neither imperfections in the apparatus nor impurities in the gas were enough to cause errors of more than three or four parts in a thousand—at any rate, in relative observations. But it does not follow that this order of accuracy should be expected in the results. For the scattering may have some influence when the measurements are pushed so far. The stopping power is proportional to the square root of the atomic weight, but the scattering to the weight itself, and though the additive law may apply strictly to either separately, it may not apply to both together. There are certainly discrepancies in the following table which are greater than experimental error, and possibly this is the reason for them. The table shows the stopping powers of certain molecules, and the calculated values on the assumption of the most probable values for the atoms which they contain. (Table IV. on p. 48.)

The stopping powers, which are given in Column II, are expressed in a novel manner, which requires explanation. They were obtained by a method, which was slightly different from the methods already described, and was adopted as capable of giving more accurate relative values. A glass bulb was connected to the ionisation chamber and thoroughly evacuated. When the determination of the ionisation curve was complete, communication was opened between the bulb and the chamber, and a certain fraction of the gas went over into the bulb. The weight of this gas was found. The relative stopping power is given by

$$\frac{\text{Molecular weight.}}{\text{Range of RaC} \times \text{weight of gas in bulb.}}$$

For the stopping power of the molecule is proportional inversely to the range, and inversely to the number of molecules traversed by the a particle. The latter quantity is proportional to the density divided by the molecular weight, and the density is proportional to the weight of the gas in the glass bulb. By this method it was rendered unnecessary to know the pressure and temperature of the gas; and what was of far more importance, there was no need to estimate the density of the gas from a knowledge of its nature, pressure, and temperature. Only two quantities were to be found, the range and the weight of the gas in the bulb. When comparing the stopping powers of different molecules, it was not even necessary to know the volume of the bulb, nor to refer to the ranges in air. Calculations and liabilities to errors of experiment were both materially reduced. The following table gives the results obtained in this way:—

TABLE IV.—*Stopping Powers: Weight Scale.*

I.	II.	III.
O_2	1553	1540
CO_2	2187	2200
CO	1437	1435
C_2H_2	1619	1618
C_2H_4	1906	1906
C_2H_6	2180	2194
CH_4	1242	1241
N_2	1444	—

The third column gives the calculated values, taking $H_2 = 288$, $C_2 = 1330$, and $O_2 = 1540$. The agreement between the two last columns is very close, and shows that the additive rule is almost exactly fulfilled. The value found for air was 1448.

The next table shows the stopping powers (expressed as usual) of these gases, both for the *a* rays of RaC, and for those of RaA: the stopping powers are not always the same for the two sets of rays, since they depend on the speed of the particle in relation to the weight of the atom:—

TABLE V.—*Stopping Powers for RaC and RaA.*

Gas.	Stopping power for RaC relative to air.	Stopping power for RaA relative to air.
O_2	1·067	1·059
CO_2	1·498	1·483
CO	0·985	0·978
C_2H_2	1·118	1·127
C_2H_4	1·349	1·374
C_2H_6	1·519	1·533
CH_4	0·860	0·880
N_2	0·989	0·992
$C_4H_{10}O$	3·437	3·471
C_5H_{12}	3·544	3·595
C_2H_5Cl	2·379	2·385

The last two gases were investigated some time before the others, and the results in their case are scarcely so accurate. The ethylene (C_2H_4) results are a little too high. We found that the density was a little more than it should be, and that quite accounts for the error. In Table IV this error does not show, as the method, by which the figures in that table were obtained, practically took this excess of density into account automatically.

The stopping power of a molecule is not altogether independent of the speed of the particle. This is shown by a comparison of the curves *A* and *B* in Fig. 21, where it is clear that the insertion of the silver film has not caused exactly the same drop throughout the whole of the curve; the slower *a* particles are a little less affected than the swifter. The stopping power increases somewhat with the speed.

The amount of the effect may be judged from the figures in the following table: they are taken from measurements of the curves of Fig. 21 and others of similar kind. The figures give the depressions in each of the various sections of the ionisation curve.

TABLE VI.—*Dependence of Stopping Power on Speed.*

	RaC.	RaA.	Emanation.	Ra.
Al	2·31	2·25	—	—
Ag	0·86	0·81	0·80	0·73
Sn	1·92	1·77	1·67	—
Au	1·63	1·46	1·36	—

In consequence the stopping power of a pair of sheets of different metals varies with the order in which the *a* particle passes through them. Let us take as an example a gold aluminium pair. The Al atom has a weight so close to that of air, that its stopping power (being defined relatively to air) is nearly independent of the speed, but the Au atom has a larger stopping effect the swifter the

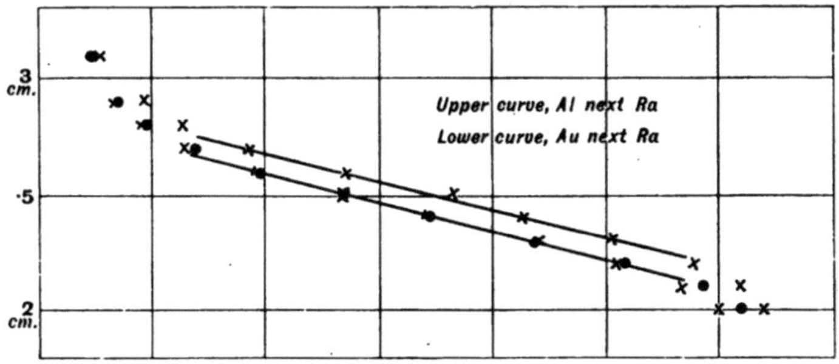

FIG. 23.

a particle. If therefore the particle goes through the gold first and the aluminium afterwards, it is in the gold when its velocity is high, and in the aluminium when the velocity is lower, and it loses more energy than when the sheets are inverted. (Bragg, *Phil. Mag.*, April, 1907). This is shown in Fig. 23; the gold sheet was equivalent to 3·4 cm. of air and the aluminium to 0·75 cm.

This is sufficient to explain—so far as it is possible to judge—an early experiment made by Mme. Curie, who found that when a pair of metal sheets was used as one of the walls of the ionisation chamber, the current was varied by the inversion of the sheets. In an experiment which I made (*loc. cit.*) for the purpose of testing the capacity of the hypothesis I found the quantities to agree very well.

Hydrogen having a lighter atom than air, its stopping power should decrease with the speed of the *α* particle; not increase, as in the cases just considered. The ionisation curve of methane shows this very well; it was found with the object of proving this point. The long range *α* particles

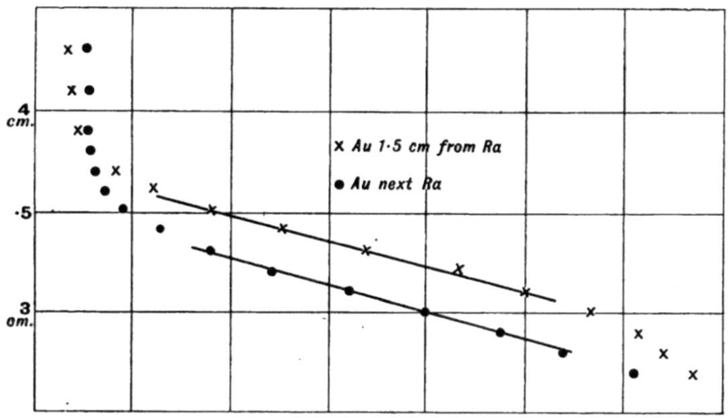

Fig. 24.

stand out more clearly in the methane curve, and the separation into four portions is generally clearer.

The drop of the ionisation curve due to the insertion of an absorbing screen, having an atomic weight different from that of air, should depend on whether it is placed near the radioactive layer, or closer to the ionisation chamber. This is shown in Fig. 24, taken from the paper already quoted (*Phil. Mag.*, April, 1907).

The question has also been very carefully investigated by T. S. Taylor (*Amer. Jour. of Science*, Sept., 1908, Oct., 1909). His work confirms the principle stated above, and he establishes the further important point that the influence of speed tends to disappear as the speed becomes higher. In other words, the stopping powers of the atoms then approach to limiting values. The behaviour of hydrogen is shown by a direct experiment. A layer of the gas is included between two very thin sheets of celloidin and inserted across the path of the a particle at various points. As we should expect, the result is exactly opposite to that of Fig. 24, where a gold sheet was used. The ionisation curve is higher when the hydrogen cell is close to the radium than when it is close to the ionisation chamber.

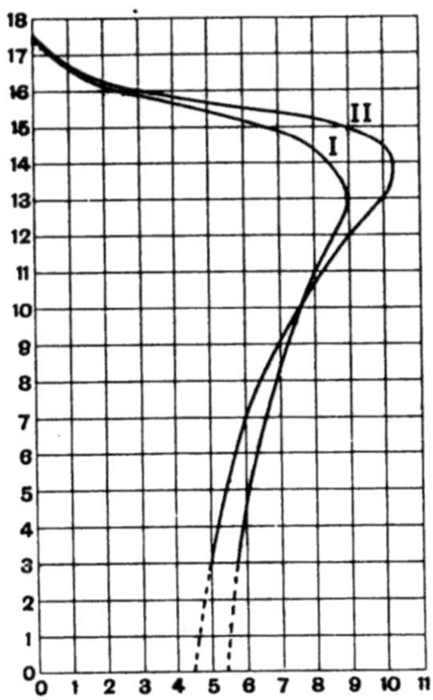

FIG. 25 (Taylor.)

I. Ionisation curve in air.
II. " " in hydrogen.

The ionisation curves in air and hydrogen show just such differences in form as the foregoing would lead us to expect. The curves shown in Fig. 25 were found by Taylor (*loc. cit.*, 2nd paper, p. 368). There is less ionisation in hydrogen than in air at high speeds, because less energy is then spent in the lighter gas. The reverse takes place at the end of the range.

Lastly, we may consider a method suitable for the determination of the range of a gas which is only obtainable in small quantity such as is insufficient to fill the apparatus described in Chapter II. In such a case I have found it convenient to use a modified form of apparatus shown in Fig. 26. Here there is no movement of the radium, but the pressure of the gas can be varied. Since the geometry of the arrangement is never disturbed, and the same stream of a rays always enters the ionisation chamber if it can reach so far, there is no need for the bundle of tubes used in the larger apparatus. Provided the radium, is of sufficient strength—and only a very small fraction of a milligramme is required—the tubular case can be made quite narrow. The pressure can be adjusted by means of a movable mercury cistern, and very accurate readings are possible.

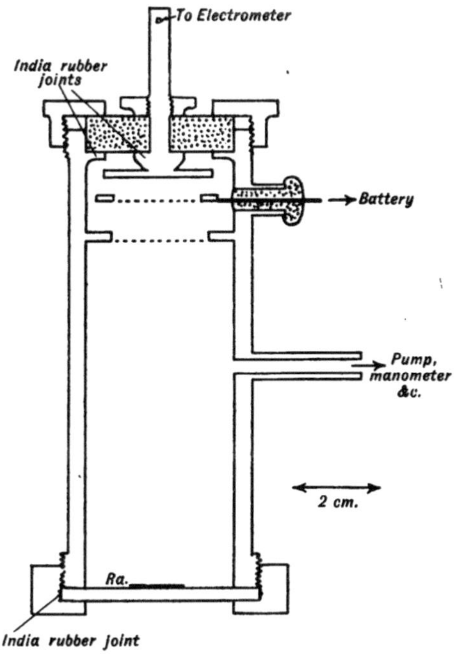

FIG. 26.—Range-finding apparatus requiring only a small quantity of gas.

The curve contains the four steps as in the former method, but the ionisation of any bundle of rays which crosses the chamber gradually diminishes to zero with the pressure. Consequently it takes the form shown in Fig. 27, which represents the result of experiments with argon and with air. By comparison of these two, we find the stopping power of argon to be 0·951 for RaC and, 0·934 for RaA. The curve also gives conveniently a measure of the relative ionisations of air and argon for a stream of α rays which is the same in both cases. We have only to compare the values of the abscissæ corresponding to the same ordinate in that part of the curve which slopes down towards the origin. This gives the relative ionisation per molecule of the two gases, a quantity which is tabulated in the column headed *ks* (p. 65). The explanation of the

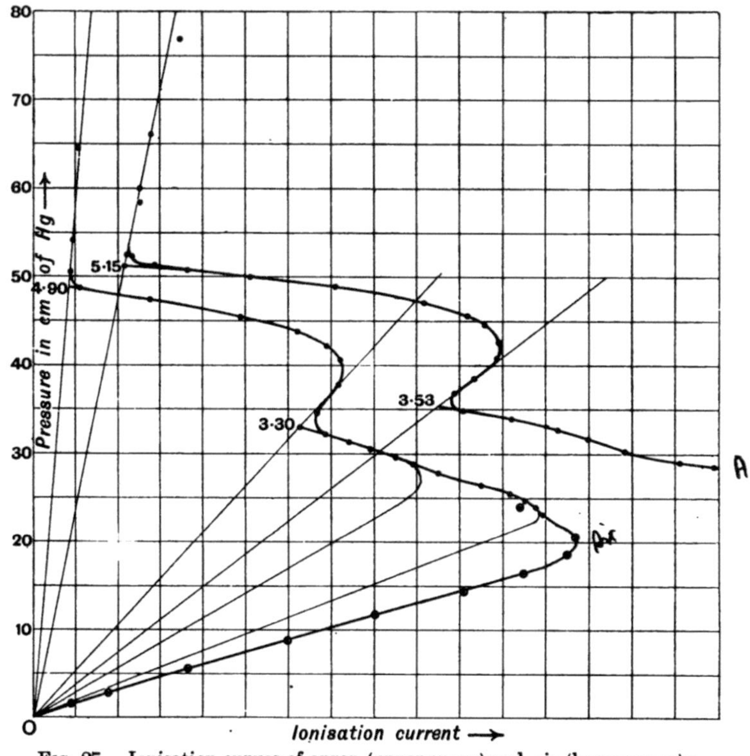

FIG. 27.—Ionisation curves of argon (upper curve) and air (lower curve); apparatus of Fig. 26.

symbol *ks* will be given presently. I have also used this apparatus to find the constants of helium. For this purpose it was necessary to put a screen over the radium so as to cut down the range of the RaC rays to such an amount that they were unable to reach the gauze of the ionisation chamber when the latter was filled with helium at normal temperature and pressure.

CHAPTER VI

THE IONISATION PRODUCED BY THE α PARTICLE IN DIFFERENT GASES

WHEN the α particle spends the whole of its energy in a gas, the ionisation produced is not the same for every gas. The determination of what may be called the total ionisation is a matter of considerable interest, and has been undertaken by several investigators. There is a special difficulty attached to it in consequence of the magnitude which the applied electric force must possess if it is to collect all the ions that are made, particularly in the case of the heavy gases. If the ionisation chamber is large enough to include the whole path of the α particle, the voltage required is practically impossible in many cases.

The difficulty may be overcome in two or three ways. The most satisfactory is to find the ionisation curve of the particles using the apparatus of Fig. 6. Here the ionisation chamber is so narrow that it is easy to apply enough electric force. The curve being drawn, its whole area is to be compared with the corresponding area when air is substituted for the gas. This method is, however, rather lengthy. In some experiments made in Adelaide (Bragg and Kleeman, *Trans. Roy. Soc. of South Australia*, Oct., 1905, and *Phil. Mag.*, April, 1906: Bragg, *Trans. Roy. Soc. of South Australia*, Jan. and Oct., 1906, *Phil. Mag.*, May, 1906 and March, 1907) assumptions were made which very much shortened the work; and though they do not now appear to have been entirely justified, it is clear that very little error was introduced by their adoption. The method may be explained thus :—

In the ionisation curve drawn in Fig. 28 the portion EA is due, as already said, to such β rays as are intercepted by the thin screen ss, Fig. 6. It may be assumed with safety that if EA is produced to meet the x-axis in D, the line EAD represents the β ray action at all distances, because the β ray action varies very slowly with the distance when it is observed directly, and is always quite small. The area between EAD,

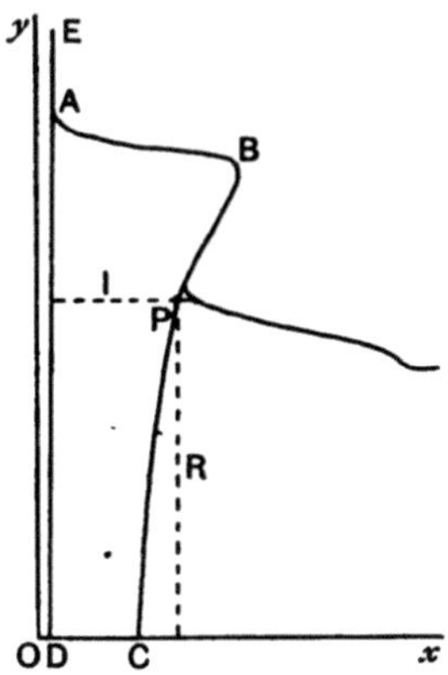

Fig. 28.

the axis of x, and the ionisation curve $EABPC$ is a measure of the whole ionisation due to the α rays of RaC. If the pressure of the gas is diminished in any proportion the ordinates of the curve are all increased and the abscissæ diminished in the same proportion, as we should expect and as experiment shows. The product of the ordinate and the abscissa of any point that can be identified is a quantity independent of the pressure of the gas.

If we take a different gas, the ionisation curve retains the same general shape, and it may be assumed without much error that the curves in two gases are similar to this extent—if they were drawn on the same diagram so that the maximum range was the same for each and the ordinates adjusted to make this so, then the one curve could be fitted to the other by increasing or diminishing all its abscissæ in a certain proportion. As a matter of fact, this is not quite the case because of the variation of stopping power with speed (see Fig. 25), but it does not appear that the error in the assumption has led to much inaccuracy. Ignoring the error, some definite point on the ionisation curve may be chosen, such as the point P in the figure where the top of the slope belonging to RaA meets the side of the curve due to RaC; and the product of the ordinate (R) and the abscissa (I) of that point may be taken to represent the whole area of the curve. This will be the case whether we compare the total ionisation in the same gas at different pressures, in which case we should find RI to be constant, or whether we compare the ionisations in different gases. This particular point on the curve is found very easily. The results in the following table show how RI is independent of the pressure so long as the gas is the same but varies with the nature of the gas.

TABLE VII.

Ethyl chloride (C_2H_5Cl).				Air.			
P. (cm. of Hg)	R. (in cm.)	I. (arbitrary units)	R × I.	P.	R.	I.	R × I.
53·8	2·87	114·0	326	75·3	4·72	54·6	257
41·0	3·78	86·0	326	57·9	6·08	43·2	262
32·5	4·83	66·6	322	46·9	7·42	34·0	252
22·4	6·92	47·6	330	38·8	9·00	28·3	254
		Mean	326			Mean	256

The ratio of the one RI product to the other may be called the specific ionisation of the gas (k): it is in this case 1·275.

A number of results obtained in this way are given in the following table, taken from the paper in the *Philosophical Magazine*, March, 1907 :—

TABLE VIII.—*Ionisation of Different Gases by the α Rays.*

	$k \times 10^2$.	$s \times 10^2$.	$ks \times 10^2$.	v.	$v/ks \times 10$.
C_6H_6	129	333	430	96·0	223
C_5H_{12}	135	359	485	117·0	242
C_2H_4	128	135	173	44·0	254
C_2H_2	126	111	140	33·0	236
$C_4H_{10}O$	132	333	440	106·0	241
C_2H_6O	123	200	246	62·0	252
CH_4O	122	143	174	42·5	244
CCl_4	132	400	528	104·0	197
$CHCl_3$	129	316	408	85·0	208
C_2H_5Cl	132	236	312	71·0	227
CH_3I	133	258	343	66·0	193
C_2H_5I	128	312	400	86·0	215
CS_2	137	218	299	62·0	207
CO_2	108	147	159	35·4	222
N_2O	105	146	153	...	...
O_2	109	105	115	24·4	212
N_2	96	96	94	...	...
H_2	100	24	24	11·0	460

In this table the description of the gas is given in the first column, the specific ionisation in the second, and the stopping power in the third. The figures in the other columns will be considered presently.

In obtaining some of these data it was necessary to heat the chamber in order to obtain a sufficient vapour density. The concordance of the results obtained for the same gas at various temperatures showed that temperature had no direct effect on the ionisation A similar result has been obtained by all who have made the test with β, γ, or X rays. A single illustration will be sufficient: the following table contains the results of several measurements of the ionisation of ethyl chloride made at different times :—

TABLE IX.—*Ionisation is Independent of Temperature.*

Volts per cm.	P.	T.°C.	R.I.	R.I. in air.	k.
1,670	23·3	14°	247	191	1·29
3,000	38·0	16°	272	204	1·33
3,000	58·0	26°	268	203	1·32
1,670	26·2	72°	211	166	1·27
1,670	31·8	60°	242	186	1·30
1,670	32·8	60°	243	186	1·30
1,670	29·4	34·5°	261	199·5	1·31
1,670	42·1	37·5°	256	198·5	1·29
				Mean	1·30

Each determination in this table required the measurement of several points on the ionisation curve on either side of *P* (Fig. 28), first in air, then in ethyl chloride and then in air again.

Laby has measured the total ionisation produced by the *a* rays of uranium in several gases (*Proc. Roy. Soc.*, 79, p. 206, Feb., 1907), using a chamber large enough to contain the full range of the rays. He found the following values of the specific ionisation (k) :—

N_2O	0·99	C_2H_2	1·27	$C_4H_{10}O$	1·29
NH_3	0·90	C_5H_{12}	1·345	CH_3Br	1·02
CO_2	1·03	C_2H_4O	1·05	C_2H_5Cl	1·18

These may be compared with corresponding values in Table II, and it will be seen that the agreement between the two sets of results is fairly good, except in the case of the gases containing heavy atoms, for which Laby's method was not suitable, since he could not apply enough electromotive force.

Kleeman found the following values, using the method of the shallow chamber (*Proc. Roy. Soc.*, 79, p. 220, March, 1907)

C_2H_4O	1·17	CH_3Br	1·32	SO_2	1·04

The last figure was calculated from his value for the relative ionisation per c.c., and the stopping power of the gas, taken to be 1·92. The method of such a calculation will be made clear by what follows immediately.

The stopping power of a gas is an inverse measure of the penetration of the *a* particle into the gas as compared with the penetration into air. It is therefore a relative but inverse measure of the number of atoms traversed by the particle. Here we interpret "traversing an atom" as meaning "crossing a space of definite volume containing the atom, the volume being taken to be of constant magnitude." Thus the free path, and the volume of the atom considered in physical chemistry, do not enter into the definition. If we multiply the total ionisation k by the stopping power s, we obtain a quantity which we may call the "specific molecular ionisation." It may be thought of as representing the probable ionisation that an *a* particle will make in crossing a definite volume containing a single molecule of the gas under consideration, in proportion to the corresponding probable ionisation in the case of an air molecule as already defined.

The value of this constant is given in the fourth column of Table VIII for the gases included in the table.

It is possible to find this constant directly. If *a* particles are projected through an ionisation chamber in which the pressure of the gas is so low that every particle gets across, the number of molecules traversed is proportional to the pressure of the gas and the number of the *a* particles. The latter being constant and the gas varied, it is easy to find the molecular ionisations.

Parr Metcalfe has measured in this way the molecular ionisations of several gases (*Phil. Mag.*, Dec., 1909). He gives the following results:—

Air		1	Propane	C_3H_8	3·05
Hydrogen	H_2	0·233	Butane	C_4H_{10}	4·02
Carbon monoxide	CO	1·00	Pentane	C_5H_{12}	4·83
Nitric oxide	NO	1·28	Helium	He	0·211
Methane	CH_4	1·10	Hydrogen chloride	HCl	1·40
Ethane	C_2H_6	2·08	Bromine	Br.	3·9

The *a* particles come from a very thin layer of uranium oxide distributed over the inner walls of the ionisation chamber.

I made some experiments in 1907 by this method, using the *a* particles of RaC, and the apparatus of Fig. 26. The value for argon was 1·245, and for helium 0·228. The former is a good result, I think; but the experimental conditions were not quite satisfactory in regard to the latter. I was unable to repeat the determination at the time, and in consequence the results have not been published previously.

There now arises a question of considerable interest. The stopping power of a molecule is very nearly the sum of the stopping powers of the individual atoms of which it is composed. But the molecular ionisations are by no means always the sum of the atomic ionisations. Kleeman has made a careful examination of this point (*Proc., Roy. Soc.*, 79, p. 220): he showed that the additive rule could be made to hold for a large proportion of the gases which he considered, but there were obvious exceptions. He assumed the following values for the atomic ionisations:—

H	0·175	N	0·47	S	1·24	Br	1·72
C	0·51	O	0·55	Cl	1·16	I	2·26

From these it would follow that the molecular ionisation (*ks*) for C_2H_5Cl should be $2 \times 0·51 + 5 \times 0·175 + 1·16 = 3·05$, whereas in Table VIII the experimental value is 3·12. So for ether the calculated value is 4·34 and the experimental 4·40: for ethylene the values are 1·73 and 1·72 respectively: for carbon bisulphide 2·99 and 2·99; and so on. But for SO_2 they are 2·34 and 2·01, for ammonia 0·99 and 0·81, while for H_2 the experimental value is 0·24, whereas it should be $2 \times 0·175 = 0·35$. Parr Metcalfe (*Phil. Mag.*, Dec., 1909) found that in the paraffins the addition of C + 2H increased the molecular ionisation by a number close to 0·92. The figures are:—

	H_2		CH_4		C_2H_6		C_3H_8		C_4H_{10}		C_5H_{12}	
	0·233		1·10		2·08		3·05		4·02		4·83	
Differences		0·86		0·98		0·97		0·97		0·81		Mean = 0·92

In this case there is some regularity in the addition. Moreover, three alcohols can be arranged so:—

	CH_4O		C_2H_6O	...	$C_4H_{10}O$	
	1·74		2·46	...	4·40	
Differences		0·72		Say 0·97 + 0·97		Mean = 0·89

The mean consequence of the addition of C + 2H is 0·89. The ethyl alcohol result is rather uncertain, because the apparatus was not working well when I found it: the alcohol seemed to have an effect on the radium, causing a liberation of the emanation, and the methyl alcohol also was troublesome. Thus the agreement between the two means 0·92 and 0·89, is quite satisfactory. Also, according to Kleeman, C + 2H should have a value 0·86, and this justifies Kleeman's values for C (0·51), and for H_2 in combination (0·35). Yet the experimental value for H_2 itself is only 0·23. We may calculate from the values for O_2, CO_2, and CO that the atomic ionisations of C and O are 0·42 and 0·55 nearly: while Kleeman's value for C is 0·51, so that here also there is a distinct want of agreement. The difference between the values for ethyl and methyl iodide is 0·57, which is much less than the value for C + 2H found above: possibly, however, these two values require re-determination.

It thus appears that while the result of the action of a molecule on an *a* particle follows an additive law very closely, the result of the action of the particle on the molecule in regard to ionisation does not do so in all cases. It is not at all clear how this arises. Possibly the *a* particle does not spend its energy directly on the process of ionisation, but on some effect which it sets up in the atom without reference to the presence of other atoms, and this effect is then left to act upon the molecule. Ionisation is to some extent a molecular phenomenon: it may even be shown to be roughly proportional to other molecular phenomena such as that of the "volume." This is shown in the fifth and sixth columns of Table VIII, the fifth giving the volume and the sixth the ratio of v to *ks*. As already mentioned, this means that the ionisation of a molecule is roughly proportional to the time the *a* particle spends within it, and this is suggestive of a possible interpretation. It is rather satisfactory to find a lack of additiveness in the molecular ionisations, because if we had nothing but purely additive effects in radioactivity, the molecule would be a mere name in the new science, and there would be no chance of linking it on to the older sciences of physics and chemistry which take so much account of molecular groupings. Hardly anything could be more interesting than to find a set of radioactive phenomena in which the additive law is closely followed, and another set nearly allied to the first in which molecular association has some little influence.

It is interesting to compare the specific *atomic* ionisations with the stopping powers: and especially to include among the former both those (headed *A* in Table X), found by Kleeman to be the best for substitution in the molecular combinations, and those (headed *B*) found by experiment with the simple gas:—

TABLE X.

I.	II. Atomic ionisation relative to air molecule.	III. Stopping power relative to air molecule.	IV. Ratio II/III.
A.			
H	0·175	0·12	1·46
C	0·51	0·425	1·20
N	0·47	0·49	0·96
O	0·55	0·53	1·03
S	1·24	0·88	1·41
Cl	1·16	0·89	1·30
Br	1·72	1·30	1·32
I	2·26	1·72	1·31
B.			
H	0·116	0·12	0·97
He	0·110	0·100	1·10
Ar	1·24	0·95	1·30

The smallest atoms give least ionisation in proportion to the energy spent: the larger atoms and molecules even though made up of small atoms yield nearly the same larger quantity.

Finally, the results so far obtained may be set out together in the following table. They have not all been obtained by the same method, and the general agreement between the values obtained by different investigators is a testimony to the exactness with which these measurements may be made. The figures to which the letter T has been attached are taken from a paper by Taylor (*Phil. Mag.*, April, 1911), and are specially interesting because the author has in every case traced out the whole ionisation curve and found its area. This is the most complete method of procedure, and Taylor's results, which seem to have been worked out with great care, must be taken as very accurate. He used polonium as the source of the α rays. It is satisfactory to find that the method which I employed in the earlier measurements gave nearly the same results, for this shows that the quicker method is trustworthy. The only serious differences are in the case of oxygen and of carbon dioxide, which was one of the first gases I investigated; in later work I obtained a value for CO_2 much closer to that of Taylor. I have included this in the table. Results marked K were found by Kleeman, those marked M by Parr Metcalfe, and I have left my own unmarked:—

TABLE XI.—*Stopping Powers* (s), *Total Ionisations* (k), *and Molecular Ionisations* (ks).

	$k \times 10^2$	$s \times 10^2$	$ks \times 10^2$
Air . .	100	100	100
H_2 . .	99 T : 100	24	22·6 Strutt . 23·3 M
N_2 . .	96 T : 96	96 {98·9 RaC, 98·2 RaA}	24 94
O_2 . . .	113 T : 103	{106·4 RaC, 105·7 RaA}	109
CO . .	1·015	{98·5 RaC, 97·6 RaA}	100 : 100 M
NO . .	...	...	128 M
CO_2 . .	101 T : 103 L : 103	{150·5 RaC, 148·8 RaA}	152
N_2O . .	99 L : 105	146	153
NH_3 . .	90 L	...	81 K
CS_2 . .	138 T : 137	218	299
SO_2 . .	103 T	...	201 K
C_2N_2 .	...	...	192 Strutt
He . .	...	20·1	21·1 M : 22·8
Ar . .	...	{95·1 RaC, 93·4 RaA}	124·5
Br . .	...	...	390 M
HBr . .	129 T	...	...
HI . .	129 T	...	...
HCl . .	129 T	...	...
CH_4 . .	118 T : 116·5	{86·0 RaC, 88·0 RaA}	110 M : 102·5
CH_4O .	122 K	...	174 K
C_2H_2 .	127 L : 126	{111·8 RaC 112·2 Em & Ra, 112·1 RaA}	140
C_2H_4 .	122	{134·9 RaC 137·9 Em, 136·9 RaA 140·5 Ra}	165
C_2H_6 .	130	{151·4 RaC, 152·6 RaA}	208 M : 197
C_3H_8 .	...	...	305 M
C_4H_{10} .	...	...	402 M
C_5H_{12} .	134·5 L : 135	{354·4 RaC, 359·5 RaA}	485
C_2H_6O .	123	200	246
$C_4H_{10}O$.	136 T : 129 L : 132	{343·7 RaC, 437·1 RaA}	440
C_6H_6 .	129	333	430
C_2H_4O .	105 L	...	214 K
CH_3I .	133 T : 133	258	343
C_2H_5I .	128	312	400
$CHCl_3$.	129	316	408
C_2H_5Cl.	129 T : 130	{237·1 RaC, 238·5 RaA}	307
CCl_4 . .	132	400	528
CH_3Br .	132 K	203	275 K

CHAPTER VII

INITIAL RECOMBINATION

When positive and negative ions are distributed through a given space, a process of combination goes on until ions of one sign only are left. Let there be p positive and n negative ions to the c.c., and suppose the distribution to be quite general so that the chance that any one of the ions is included in any given element of volume is independent of whether there are others there or not. Then the number of recombinations taking place in an element of time δt is $\alpha pn\delta t$ where α is a constant called the "coefficient of recombination." It has been the subject of measurement by Rutherford, Townsend, McClung, Langevin and others.

But when α particles are fired across a gas the initial distribution of ions is by no means of the character supposed above. The geometrical relations of the ions to each other are such that there is generally a very rapid recombination as long as this special distribution lasts: and the process may be termed "initial recombination."

This effect manifested itself in certain experiments made by Kleeman and myself, to which reference has already been made. Varying the electric force applied to the ionisation chamber in order to make sure that all the ions were being collected, we found that far larger forces were required than observers had found necessary hitherto, but we did not realise at first that we had stumbled upon a characteristic property of α rays. It next appeared that when the ionisation was not too large, the loss of ions through recombination, when the electric force was insufficient, was independent of the density of the ionisation: a fact in entire opposition to what had usually been observed. And again we found that the dimensions of the ionisation chamber had very little influence, which was also contrary to what we had expected on the ordinary theory. For if a field of force is applied across a chamber in which ionisation is being produced the positives and negatives move past each other, and though the number of recombinations taking place in each second is independent of the presence of the field—the velocities due to the latter being small compared with the velocities of thermal agitation—yet the field limits the time during which it is possible for meetings to occur. Experiments showed, however, that it made very little difference whether the depth of the chamber was 3, 6, or 9 mm. so long as the electric force (volts per cm.) was kept constant. The results of the experiment are shown in Fig. 29, which has been redrawn from the figures given in the original paper (*Trans. Roy. Soc. of South Australia*, October, 1905; *Phil. Mag.*, April, 1906). The experiment has been repeated and the result confirmed by Moulin (*Recherches sur l'ionisation produite par les rayons* α, February, 1910. Paris: Gauthier-Villars), and by Wheelock (*Amer. Journal of Science*, October, 1910).

These experiments show that recombination of ions is taking place in such a way that it cannot be due to the general recombination of positive and negative ions distributed entirely at random through the gas. Since ions tend to diffuse and to assume a general distribution, the recombination must take place before they can do so, and must be the result of some special distribution of the ions when first produced. This may also be seen from a numerical example. In one of the experiments described, the number of ions falling on each sq. cm. of the ionisation chamber in a second was about $1{\cdot}2\times10^{6}$. The width of the chamber was 4 mm., in which distance it appears from Rutherford's work that each *a* particle would

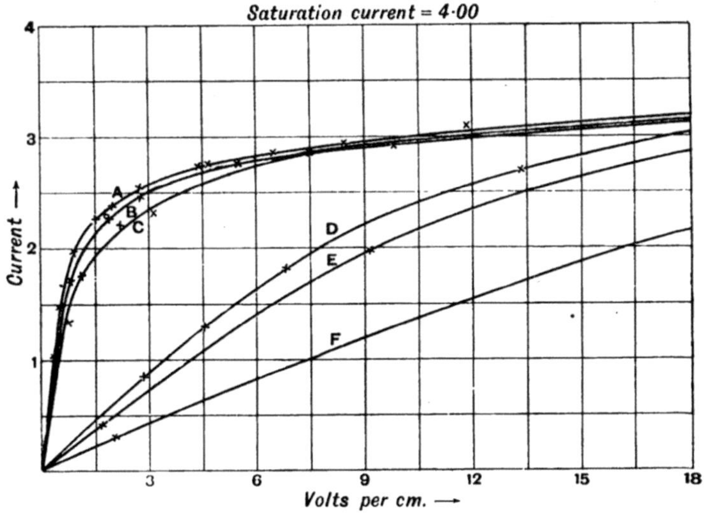

FIG. 29.—Curves showing the relation between the potential gradient and the ionisation current in air in various cases.

Curve	A.	B.	C.	D.	E.	F.
Width of ionisation chamber in mm.	3	6	9	3	5	10
Order of saturation current in amperes per cm.[2] of electrode	10^{-12}			10^{-10}		

Curves *D*, *E*, *F*, are drawn from results obtained by Retschinsky, *Ann. der Phys.*, 1905, p. 518. They show effects due to general recombination in presence of the large ionisation. The other three curves show recombination which is almost entirely initial.

create some 40,000 ions. Hence about thirty *a* particles crossed the chamber per sq. cm. per sec. With an impressed force of only 25 volts per cm. the speed of the ion would be about 40 cm. per sec., and it is therefore clear that the ions made by any one *a* particle were usually swept away before another α particle fell on the same sq. cm. of plate. As a matter of fact, if they all came at once the distances between the separate tracks were such that no ions could cross from one track to another in a fortieth of a second. Recombination was only possible between the positives and negatives of the same column.

Recombination cannot often occur between any ion and another of the opposite sign belonging to the same column at any distance comparable with the width of the ionisation chamber, for the width of the chamber has practically no influence on the result. The electric force is along the column in the experiments illustrated in Fig. 29, and the negatives and positives therefore slide past each other in oppositely directed streams. The longer the column the greater the number of positives passed by any one negative, and if this kind of recombination took place frequently the length of the column, that is to say, the width of the ionisation chamber, would have much influence: and it has not. We are forced to the conclusion, therefore, that the recombination which we are considering takes place between any ion and its immediate neighbours of opposite sign in the same column.

In an experiment described in the original paper (*Phil. Mag.*, April, 1906) an attempt was made to discover any relation that might exist between the direction of the applied field and the direction in which the electrons were projected from the atoms; for it was conceivable that there might be some such definite direction. The α particles were shot across the ionisation chamber in a slanting direction so as to make an angle with the lines of force, and we looked for any alteration in the electric force required to extract a given fraction of the whole number of ions produced. The results were "practically negative." This was an unfortunate experiment. It turns out that there is actually a variation of the force required; not large in the particular case we investigated, but we might well have found it. Moulin obtained the positive result (*loc. cit.*) and examined the whole question very carefully and fully. The following table contains examples of his results; they are selected from the many clear illustrations of the phenomenon which he gives in his paper. In the first column is put the electric force in volts per cm., in the second the ratio of the current at this voltage to the maximum current where the a rays are perpendicular to the field, and in the third the same ratio where they are parallel.

TABLE XII.—*Initial Recombination Influenced by the Direction of the Field.*

Volts per cm.	Rays perpendicular to the field.	Rays parallel to the field.
3·1	0·650	0 605
5·3	0·730	0·670
7·5	0·790	0·705
11·9	0·850	0·735
22·8	0·920	0·785
67	0·965	0·855
220	0·995	0·935
660	1·000	0·975
2,900	1·000	1·000

These figures show that it is distinctly easier to obtain the full current in the gas when the electric force is applied at right angles to the path of the a particle, and we conclude therefore, with Moulin, that recombination is apt to take place between an ion and its neighbours in the same column of ions which the a particle produces. The field which is applied along the track of the particle draws them past one another, that which is applied at right angles draws the positives and negatives away from each other. The ionisation along the track of the a particle is very dense, and no doubt there is a special likelihood of combination between immediate neighbours, which is partially counteracted by the imposition of a field at right angles to the path.

In the ordinary form of chamber the α particles do not for the most part cross the chamber exactly at right angles to the walls, and they therefore make small angles with the direction of the field which is exactly normal. An interesting experiment has been made by Wheeler (*loc. cit.*, p. 246), using an ionisation chamber of which the upper wall was part of a sphere of 3 cm. radius and the lower (of wire gauze) was part of a concentric sphere of 2·6 cm. radius. The radioactive source was placed at the common centre, and therefore both the lines of force and the tracks of the

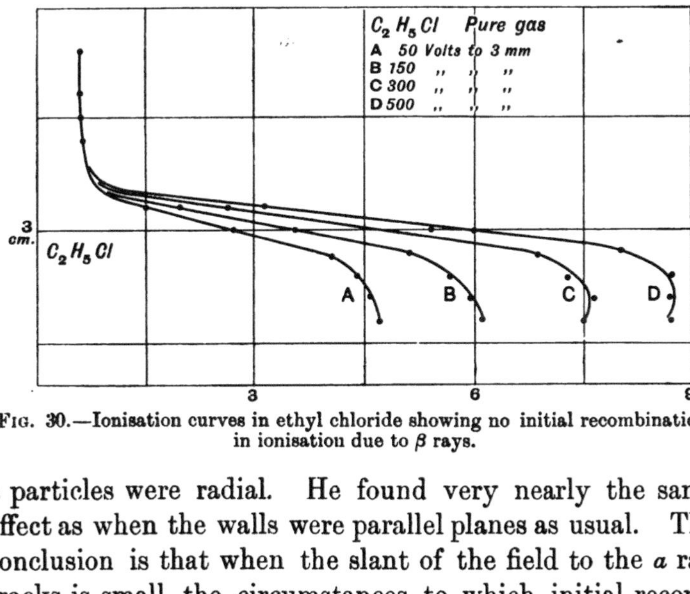

FIG. 30.—Ionisation curves in ethyl chloride showing no initial recombination in ionisation due to β rays.

α particles were radial. He found very nearly the same effect as when the walls were parallel planes as usual. The conclusion is that when the slant of the field to the α ray tracks is small, the circumstances to which initial recombination is due are over before the small normal component of the field can clear the two sets of ions from each other.

There are other interesting experimental results. In the first place, there is no appreciable initial recombination when ionisation is brought about by β, γ, or X rays. Consider the curves in Fig. 30, which are upper portions of the ionisation curves in pure ethyl chloride for four different values of the electric force (*Trans. Roy Soc. of South Australia*, Oct., 1906; *Phil. Mag.*, March, 1907). The existence of initial recombination is clearly shown in the increase of α ray ionisation with increase of the electric force applied. But the portion of the curves which represents β ray action is not shifted by changes in the electric force. Kleeman also has shown this to be true (*Phil. Mag.*, Oct., 1906) and extended the same principle to the cases of X rays and γ rays.

Again, initial recombination is greater towards the end of the path of the α particle. This was shown by Kleeman (*Phil. Mag.*, Oct., 1906) and by myself (*Trans. Roy. Soc.*

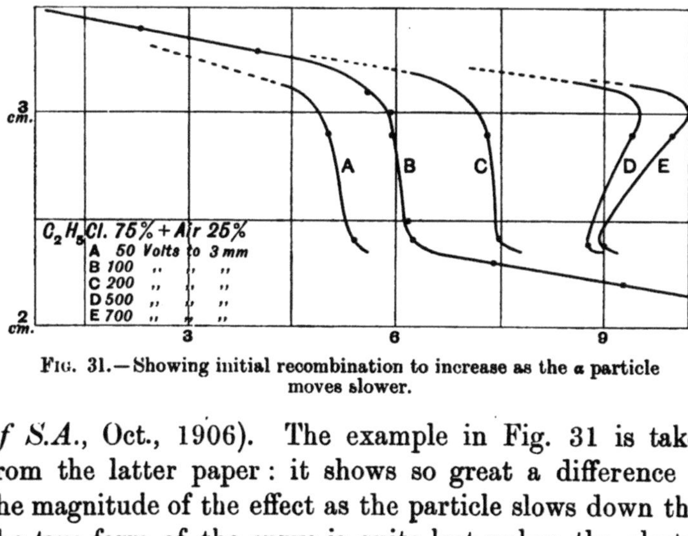

FIG. 31.—Showing initial recombination to increase as the α particle moves slower.

of S.A., Oct., 1906). The example in Fig. 31 is taken from the latter paper: it shows so great a difference in the magnitude of the effect as the particle slows down that the true form of the curve is quite lost unless the electric force is large. No doubt this is due in part to the increase in the density of the ionisation towards the end of the path. Moulin has suggested (*loc. cit.*, p. 44) a test of this hypothesis. He supposes the time during which initial recombination may occur to be the same at all points of the path. If that time is T and if a is the rate of recombination we shall have as usual $\delta n = -an^2\delta t$ where n is the number of ions of either sign remaining after time T has elapsed, and N was the original number. From this we find that $1/n - 1/N = aT$: that is to say, $(N-n)/Nn$ is a constant. Moulin found this to be nearly true in the case of air: but it is easily seen that if the formula is applied to the ethyl chloride curves of Fig. 31 it fails. It has been suggested that the speed of the electron ejected from the atom in the act of ionisation diminishes with the speed of the a particle, and therefore that the electron is more difficult to remove from the neighbourhood of its parent atom when the a particle moves more slowly. We have no knowledge of any such relation between the velocities of the electron and the a particle. According to the most recent experiments (Campbell, *Phil. Mag.*, April, 1912), the electron moves far slower than the particle, so slowly indeed that its velocity has not been measured as yet: there might be a connection between the velocities but there is no obvious reason for one.

Initial recombination diminishes rapidly with the pressure of the gas. This is illustrated in Fig. 32, taken from the original paper. If ions escape recombination by diffusing away from the column in which they are found, we should expect an effect of this kind. No doubt a similar explanation can be given of the diminution of initial recombination with rise of temperature, an effect illustrated by the figures in the following table (*Phil. Mag.*, March, 1907):—

TABLE XIII.—*Effect of Temperature on Initial Recombination.*

CO_2.	Ionisation at 1,000 volts per cm.	Ionisation at 333 volts per cm.	Ionisation at 166 volts per cm.
(*a*) 651 mm. 20° C.	100·0	95·0	90·2
(*b*) 760 mm. 72° C.	100·0	96·8	94·0

In the two cases compared the variation of temperature is so compensated by variation of pressure that the density remains the same.

Finally we may consider the most striking of all the

initial recombination effects, the great dependence upon the nature of the gas. In Fig. 32 we may observe the difference in the magnitude of the effect in air and in ethyl chloride, and in Fig. 33. I have shown the result of experiments made at different times with some other gases. In general the effect increases with the complexity of the

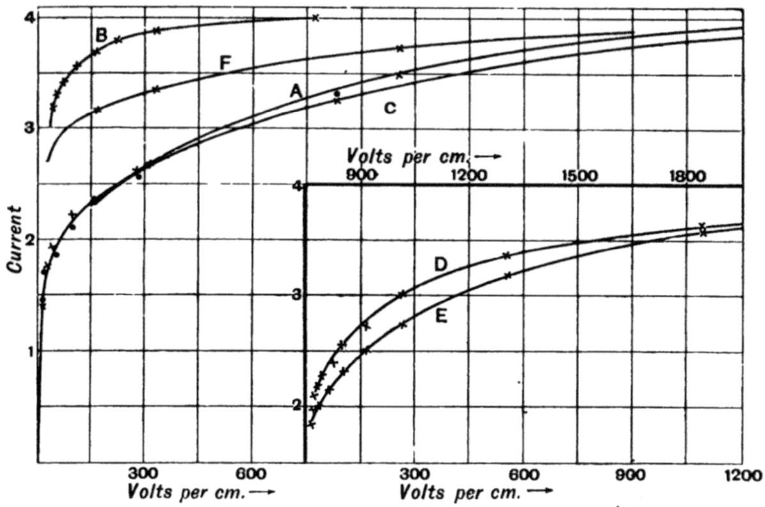

FIG. 32.—Curves showing the relation between the potential gradient and the ionisation current in ethyl chloride and air.

A. Ethyl chloride: width of chamber, 3 mm.; pressure, 56 cm.; saturation current, 2×10^{-13} amp.

B. Air: width of chamber, 10 mm.; pressure, 76 cm.; saturation current, 2×10^{-13} amp.

C. Ethyl chloride and air: pressure 76 cm.; results for 2 mm. chamber denoted by crosses; for 4 mm. chamber by circles.

D and E. Ethyl chloride and air: saturation current in E = six times the saturation current in D.

F. Ethyl chloride: width of chamber, 3 mm.; pressure, 36 cm.; saturation current, 2×10^{-13} amp.

molecule. But it is also larger in argon than in air, and in carbon monoxide it is quite small. The densities, molecular ionisations, stopping powers and diffusion coefficients of carbon monoxide, air and ethylene are much the same; yet there are marked differences in their initial recombination effects.

It may be that the effect depends in part on the fact

that by the laws of chance more than one of the atoms of the same molecule may be ionised by the same *a* particle, for the effect of an atom on an *a* particle is not altered by the presence of neighbouring atoms. Perhaps the molecule still hangs together if this happens; one would rather expect that it does not. But in either case there will be special chances of recombination until the little gathering is dispersed.

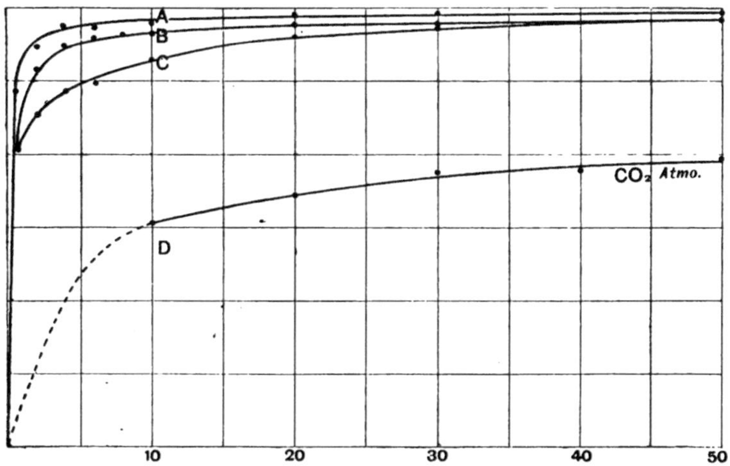

FIG. 33.—Saturation curves in A, carbon monoxide; B, ethylene; C, air; D, carbon dioxide; all at atmospheric pressure. All are drawn to the same maximum ionisation. The abscissæ represent volts per cm.

Altogether the precise cause of initial recombination cannot be said to have been explained fully as yet. Further work is certainly required, particularly in the way of comparing the effect in different gases so that relations may be found, if possible, between initial recombination and other properties of the gases.

CHAPTER VIII

THE β RAY: AND THE LAW OF ITS SCATTERING

The β rays emitted by radioactive substances and the cathode rays of the vacuum tube both consist, as is well known, of electrons moving at a high rate of speed. The β rays emitted by RaC have initially a velocity little less than that of light: the majority of the β radiations have velocities somewhat smaller than those of RaC. Cathode rays may have a great variety of velocity, being usually excited by the applied electromotive force to a speed of about 10^9 to 10^{10} cm. per second. Even electrons moving with still less speed, down to somewhere about 10^8 cm. per second, may be classed as β rays.

For all these rays have the common property of being able to ionise a gas through which they pass. The fastest have of course the greatest energy to spend, and they can penetrate two or three metres of air at ordinary temperature and pressure, ionising some of the molecules through which they pass: the slowest are barely able to penetrate at all, and the limit of speed below which they are unable to show their presence by their ionising effects is also the limit to their definition as β rays.

There is no need to discuss here the determinations of the mass, charge and velocity of the various electrons from various sources. I may pass on at once to consider the effects attending their passage through matter. They may be taken in much the same order as that which I have adopted in considering the α rays; and I shall be glad to follow this plan because it is one of my main objects to show that the phenomena of the α and the

β rays, and indeed also of the γ and X rays, run parallel to each other and may be classified according to the same system.

The means by which the β rays are discovered to us are the same as in the case of the α rays. They, like the latter, can ionise a gas, act on a photographic plate, and cause a screen to phosphoresce. It is therefore possible to map out the paths which the β rays take as a stream, but the effect of a simple β particle has been too feeble to follow until quite recently. Just as Rutherford and Geiger's scintillation method enabled them to follow the movement of the single α ray, so now C. T. R. Wilson has made a similar advance in the case of the β ray. By a modification of his beautiful fog experiment he is able to render the track of the single β particle visible to the eye : his success is, however, so recent that few results are yet available.

In the case of the α rays we have seen that the phenomena can be grouped about two main points—one, the loss of energy in passing through matter; the other, the chance of scattering by encounter with the atom. We may consider the β ray phenomena in exactly the same fashion. But we must add a third, the chance of the β ray disappearing entirely in favour of an X or γ ray which then makes its appearance. As compared with the α ray, the β ray is far more liable to scattering; it is the most obvious of its phenomena. In the case of the α ray it is the loss of energy during penetration that is the chief thing to measure.

Let us therefore first consider the general effects of scattering. When a pencil of β rays is allowed to fall upon a plate of any substance similar rays are found to be spreading in all directions from the place of impact. If the plate is not so thick that the rays cannot get through it, this so-called "secondary β radiation" occurs on both sides of the plate, and is more or less distinguishable from rays which can be looked on as a continuation of the incident stream. It is natural to expect that β rays should be scattered in passing through matter. The α rays, though enormously more massive, are occasionally deflected, especially when their speed is small. The α ray has a positive charge of twice the unit of electricity, and the β ray the single negative charge: but since the mass of the latter is some four thousand times less than the former, deflections must occur far more easily.

The conception of the scattering of a stream of electrons is an old one. Lenard investigated the passage of the cathode rays through very thin aluminium windows in the vacuum tube and drew diagrams of the effects observed in the air or other gases outside. He discussed the transmission of the rays through the window, and the "diffusion of the stream." (*Wied. Ann.*, li, 1894). Examples of these diagrams have been given already (Fig. 1). Many of Becquerel's investigations of the photographic effect of secondary β rays show the same scattering process: these being of course in relation to swifter electrons than were considered by Lenard. We shall have need to consider some of Becquerel's results at a later stage.

The β radiations which spread away in this fashion from the place where the primary rays encounter matter are found to have approximately the same power of penetrating matter as the electrons in the original stream: a little less but never greater. This also was shown by Becquerel, whose photographs are most illuminating in relation to all the points we are considering. Eve proved the truth of this principle also (*Phil. Mag.*, Dec., 1904), showing that some substances returned a rather more penetrating secondary radiation than others, though the differences were not important. McClelland has also examined the point in some detail, and the following table will give an idea of the quantitative relations (*Proc. Roy. Soc.*, lxxx,

p. 501, 1908). The figures show the fraction of the secondary radiation from various metals which is able to get through an absorbing layer of three sheets of tinfoil.

TABLE XIV.—*Penetrating Power of Secondary β Rays.*

Substance emitting secondary rays.	Intensity after passing through tinfoil. / Intensity before passing through tinfoil.	
	Incident rays, normal. Secondary rays, 15° to normal.	Incident rays, 60° to normal. Secondary rays, −60° to normal.
Pb	0·63	0·74
Pt	0·63	0·74
Sn	0·60	—
Ag	0·59	—
Cu	0·57	0·74
Al	0·50	0·70

The corresponding ratio for the primary rays was 0·75. These secondary rays all emerged from the side of the plate on which the primary rays were incident.

In the case of the figures in the second column, the primary rays were incident normally as along P_1A, and the secondary measured were those which proceeded along AS_1 (Fig. 34). For the third column the primary proceeded along P_2A and the secondary along AS_2, thus making an angle of only 60° with the primary. In the latter case, the secondary were nearly as penetrating as the primary rays, in the former they were distinctly less so, and less the smaller the atomic weight of the material of the plate.

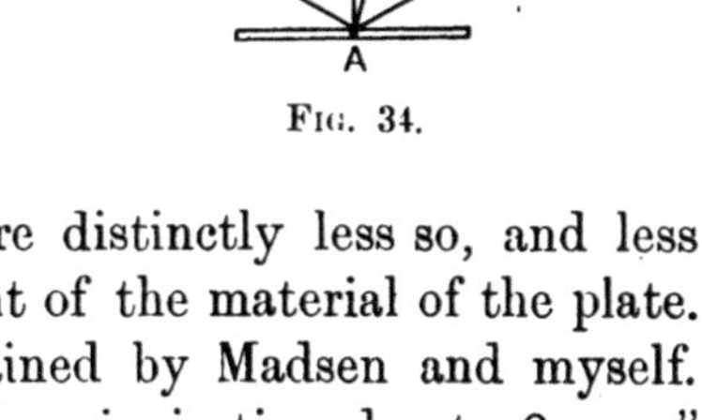

FIG. 34.

Similar results were obtained by Madsen and myself. ("The quality of the secondary ionisation due to β rays." *Trans. Roy. Soc. of South Australia*, Oct. 1, 1907.)

It is to be remembered that the penetration of an electron varies very greatly with its speed. Cathode rays, which have a speed of 10^{10} cm. per second, only penetrate two or three millimetres of air; while the β rays, which have three times the speed, can go perhaps a thousand times as far. Consequently the differences in penetration which are shown in the table on p. 79 represent very small differences in velocity. The differences are all of the nature of diminutions; and on any theory which ascribes these effects to pure scattering this is to be expected, since an electron cannot gain sensibly in velocity through encounter with an atom.

It has also been shown that when cathode rays fall upon a plate there is an issue of similar rays from the place of impact, consisting of electrons moving with a variety of speeds ranging from that of the primary downwards.

All these secondary radiations may well be electrons of the original or cathode stream diverted by the atoms upon which they have fallen. It is indeed conceivable that a β ray might impart to an electron belonging to some atom such a speed that it then became a β ray, the energy of the primary being shared between the two. It would be very difficult to decide on the present evidence whether such a thing ever occurs: and perhaps it is unnecessary to force a decision as yet.

A point of far greater importance arises when we consider the source of energy of the secondary radiations. It is simplest to suppose that the energy is all drawn from the primary; so that even if there is uncertainty as to whether the electrons in the secondary are identical with such as were originally in the primary, nevertheless, the energy in the secondary all comes from the primary, and the energy of the secondary is simply primary energy scattered.

This point is fundamental, because if there is any other source of energy, if for example, the atom is so disturbed by the entry of a β ray as to expel a new β ray, drawing on its own energy to do so, we have a physical phenomenon of extreme interest and importance. Now it is true that radioactive atoms expel β rays, but on the other hand they cannot be prompted to do so. Not even radiation by α and β particles from other radioactive substances, or other portions of the same substance, can hasten or delay their own emission, and generally the energy of radioactivity is not subject to any known control. A β ray is indeed emitted sometimes when a γ ray meets an atom, but it will be seen later that there is no need to suppose the atom to be prompted to radioactivity, and that there are many reasons against such a hypothesis. It is far more likely that the energy comes from the γ ray.

No doubt the example of the radioactive effects, and especially the discovery of induced radioactivity, made it once possible to put forward such a hypothesis with little hesitation. Atoms were found to radiate energy in comparatively large quantities, and it was natural to suppose their activity might also be stimulated. The very name, "induced radioactivity," is a witness to the existence of the idea. There is nothing which could warrant us in making such an assumption now. In any investigation of the properties of β rays it is therefore better to avoid the introduction of an assumption of this magnitude and, since all the facts can be regarded as due to a simple scattering of the primary rays, to take the latter course. There is indeed no need at present to go further than the still simpler and more limited hypothesis that the electrons of the secondary radiation were all at one time included in the primary. We are, in fact, investigating nothing more complicated than the behaviour of a stream of swift electrons playing upon atoms of matter and being scattered in consequence.

Another fact which tends to simplicity is the additive rule in regard to the action of a molecule upon a β ray. Just as in the case of the α particle, the effect of the molecule is the sum of the effects of the individual atoms of which it is composed. We shall see indeed that the parallel effect holds in the case of γ and X rays also, and is a general principle of radioactivity.

The point has been proved by several investigators. McClelland showed that the β rays returned from a plate of known molecular constitution could be calculated from a knowledge of the amount returned by plates composed of the various atoms of the molecule (McClelland and Hackett, *Royal Dublin Society's Trans.*, April, 1906. *Cf.* also Crowther *Phil. Mag.*, Oct., 1906). Again, H. W. Schmidt showed that the percentage of β ray energy absorbed in passing through a compound could be calculated from a knowledge of the loss in passing through the several elements of the compound (*Phys. Zeit.*, April 1, 1910). This investigation was specially undertaken in order to re-examine the truth of the additive law, upon which some doubt had been thrown. Borodowsky carried out an independent inquiry at the same time with the same end in view, and came to the same conclusion (*Phil. Mag.*, April, 1910). He found, for example, that the absorption of a CS_2 molecule could be calculated within one per cent. from separate measurements of the absorptions of carbon and sulphur. The β ray is never under the influence of more than one atom at a time.

We now come to a very interesting side of the question. A β ray being supposed to be projected against a plate of any given material, we ask what likelihood there is that it will be turned aside in going a given distance, what will be the probable angle through which it will be turned when it meets an atom, how that angle depends on the nature of the atom, and on the speed of the ray, and so on.

Now if the β particle suffered as many encounters, that is to say occasions of deflection, as an atom would in going the same distance, the problem would be almost hopeless. But there is quite good evidence that the β particle pierces great numbers of atoms without deflection, that the number of encounters which the secondary β rays have had is quite small generally, and that with experimental arrangements which are easy to make one encounter only has been the cause of the observed deflection of th secondary β ray.

That the β particle can pass through many atoms without deflection is very clearly shown by the original experiments of Becquerel, of which a reproduction is given in Figs. 36 and 37. An illustration of the arrangements is given in Fig. 35. These figures are taken from Mme. Curie's *Traité de Radioactivité,* Vol. ii. Plate v.

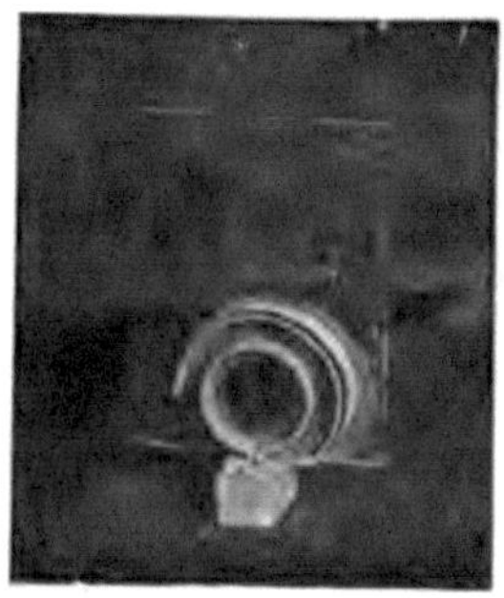

FIG. 35.

Two little circular screens, pierced with holes as shown, rest against a photographic plate. The β rays from a little radium at the bottom of the apparatus swing round in circles under the influence of the magnetic field which is applied normally to the plate, and act on the plate as they graze it. The jets of β rays issuing from the holes in the outside screen lie on circles which pass through the radium, the least speedy lying on circles with the smallest radii. A screen is placed over the outer holes, but if it is not too thick it does not destroy the line of movement of, at any rate, the faster rays. When the screen consists of aluminium 0·1 mm. thick, the faster pencils go through with very little diffusion. The slower hardly get through, and if they do

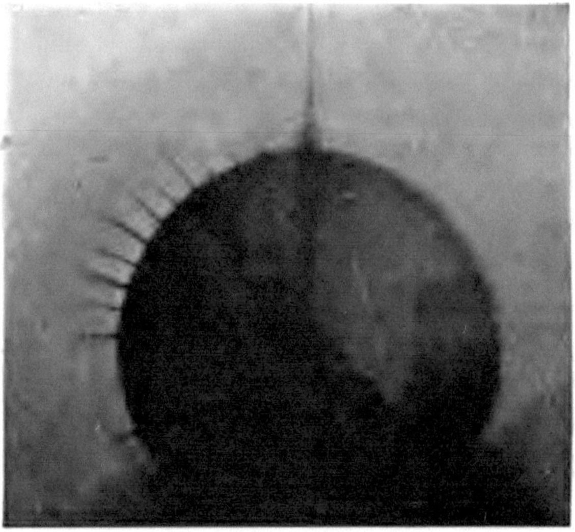

FIG. 36.—In this case the β rays have traversed a layer of aluminium 0·1 mm. thick placed over the outer holes as shown in Fig. 35.

FIG. 37.—In this case a layer of paraffin 8 mm. thick is placed over the outer holes.

are more diffused. Even when the screen is a layer of paraffin 2 mm. thick, the faster get through, and though they are more diffused in this case they still have a definite line of movement in continuance of the old. When the layer of paraffin is 8 mm. thick, the rapid rays can no longer cross it and stop at a depth of 2 mm.; the less rapid penetrate in the order of their velocity and come to a sudden end, which is marked by a maximum of impression upon the plate (Curie, p. 53). This is all very like the effects of the *a* ray.

Madsen describes a striking experiment which supports

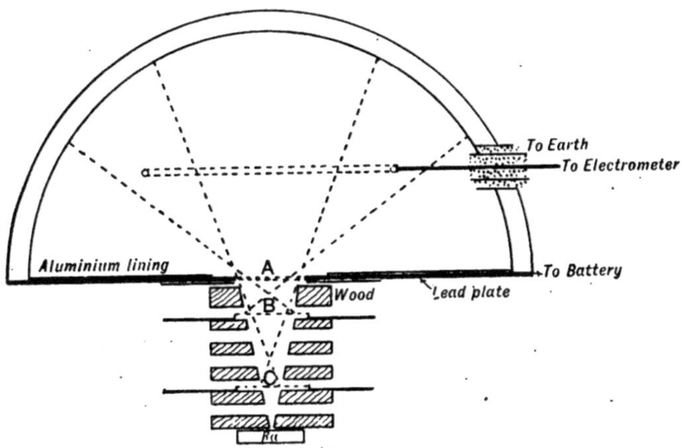

FIG. 38.

the same idea (*Phil. Mag.*, Dec., 1909). His apparatus is shown in Fig. 38. A hemispherical ionisation chamber is made of wood lined with aluminium foil: the purpose of this arrangement is to avoid scattering at the walls of the chamber as much as possible. The radium is placed at the bottom of a conical hole cut in a block of wood as shown, and at *A*, *B*, and *C* are slides by which absorbing screens may be introduced. The hemispherical form gives an equal path in the chamber to every ray that gets through, and an equal record of its presence there.

When the screen is at *A*, all the β rays which get through it make their way into the chamber: when it is

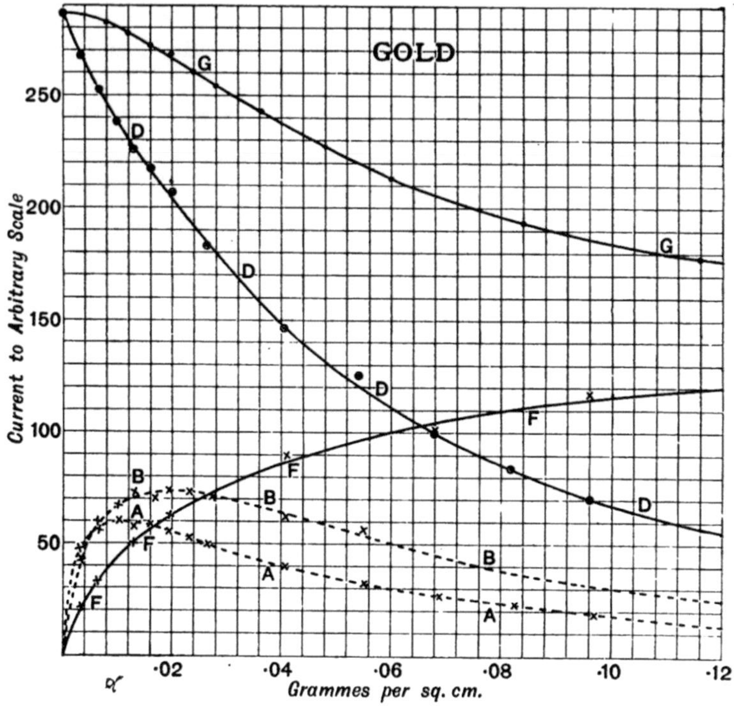

FIG. 39.—Cathode ray current from

at B only such rays get in as emerge from the screen at less than about 60° to the normal: and when it is at C, the entering rays are limited to those that make an angle of less than 30° to the normal. By subtracting the ionisation in the C position from the ionisation in the B position for all thicknesses of screen, and then again the B values from the A values, Madsen obtained measures of (a) such rays as are scattered more than 30° and less than 60°, (b) such as are scattered more than 60° and less than 90°. Let us call these quantities the less scattered and the more scattered radiations; they are represented by the ordinates of the curves A and B in the figure, which are taken from Madsen's paper. Now it appears that for small thicknesses of screen the ratio of the more scattered to the less scattered is constant, and this holds up to a thick-

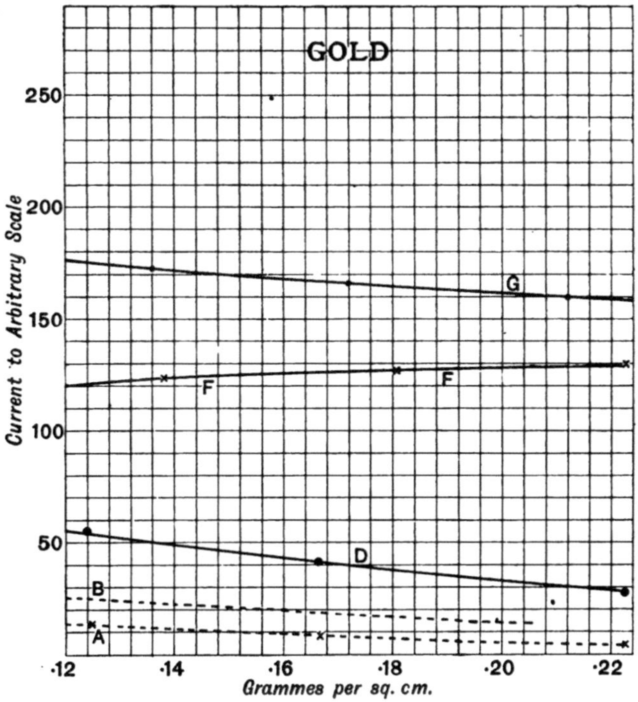

gold screens of varying weight.

ness of 0·04 mm. of Al, and proportionately less for gold. This is what we should expect if those particles which have been turned through more than 30° from their original direction have as a rule suffered no more than one deflection. For then, if a screen of 0·02 mm. of Al has been used and the quantities of more and less scattered radiation have been measured, the addition of a second screen of the same thickness will deflect from the primary pencil further quantities in the same proportion and will not interfere appreciably with the ratio between the quantities already deflected.

We may draw the same conclusion from the way in which such curves as *A*, *B*, and *F* (Fig. 39) approach the origin. These were obtained by Madsen by the use of absorbing screens of varying thickness down to the thinnest that could be obtained. They make a finite angle with the axis of x, and are of the type of the curve P in Fig. 41, not of Q. There is no sign even for the thinnest leaf used (about 0·0001 cm. Al) of any point of inflexion on the curve, so that the tangent to the curve at the origin is to all appearances more inclined to the

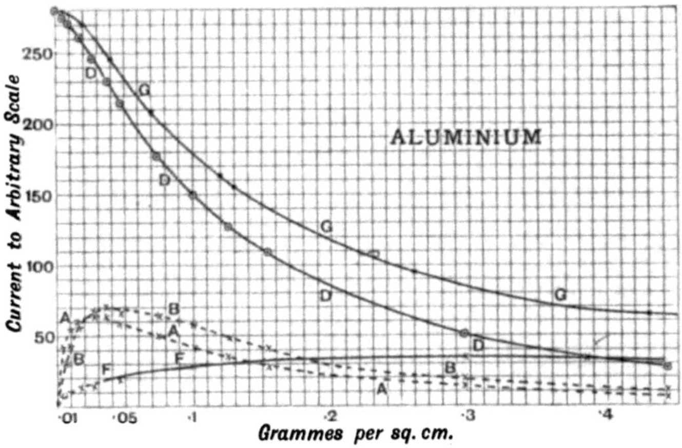

FIG. 40.—Cathode ray current from aluminium screens of varying weight.

axis of x than the tangent at any other point. This means that for the very thinnest films the fraction of the rays scattered bears a definite ratio to the mass per sq. cm. of the plate. The slopes of such curves at the origin are relative measures of the rays scattered by sheets so thin that no β particle can have had more than one encounter in crossing the sheet. That is to say, they represent the results of the encounter of the single β particle with the single atom, showing the probable chance of deflection in each direction for each kind of atom investigated.

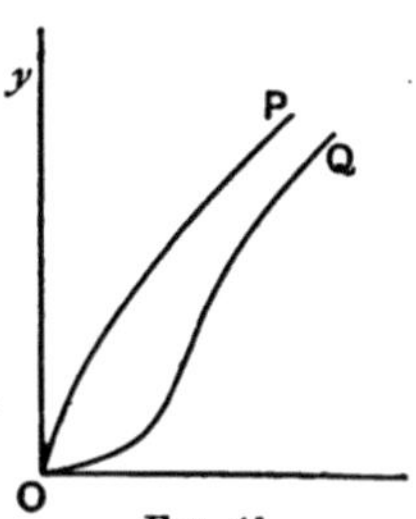

FIG. 41.

The results for any given case may conveniently be represented by drawing radii from a point representing the atom, each of a length proportional to the chance of deflection into the direction in which it is drawn. The extremities of these radii will lie on a surface, the form of which will represent the scattering effect graphically, and the form will vary with the nature of the atoms and the speed of the β ray. The surface is one of revolution, the axis of which is the original direction of the β ray. A section through the axis will therefore express the facts completely, and such a section may be called the "deflection oval." It is a proper object of experiment to determine the oval in all possible cases, for as the result we gain knowledge which will be of service in the inquiry into the nature of the atom. Moreover, when we have learnt the probable results of an encounter between a β ray of given velocity and an atom of given nature, so far as deflection goes, we are so much nearer to being able to calculate the results when a pencil of β rays falls on a plate which is an aggregate of atoms. When the simpler problem has been solved, we shall be in a position to approach the more complex—and this is the right way to proceed.

It seems now to be clear that in much work on the β rays, the opposite procedure has been followed and the complex problem has been the first to be experimentally attacked. Measurements have been made on the "reflection" from, or "absorption" by, thick plates, in which every source of complexity occurs, variety in the speed of β rays, accumulation of scattering at multitudinous encounters, loss of speed, and continual alteration of scattering coefficients and other constants and so forth. A theory is required as a guiding line through the maze, and it is of necessity an approximate one and of limited application. All that the theory can do is to arrive by rough analysis at the probable consequences of the encounter between the single β ray and the single atom, and this we see can be obtained by direct experiment, provided of course that we are right in the interpretation of Madsen's results.

We have yet to take into account the loss of speed of the particle in passing through matter, an effect of great importance in experiments with thick plates, and we shall see presently how we may investigate this point without interference from the phenomena of scattering, just as Madsen's experiment gives the scattering without interference from the effects of loss of speed. The two effects, together with the chance of conversion into X or γ ray form, fill up the history of the particle.

There is yet much information to be desired on the subject of the form of the deflection oval. We know, however, that the lighter the atom the more eccentric the oval, the heavier atom being more capable of swinging round the β ray which tries to pass by. This is in agreement with the results of Eve, McClelland, and others who have allowed β rays to fall upon thick plates of various substances and have found the returned radiation to increase with the atomic weight of the material of the plate.

Density of material is a matter of no consequence in such experiments, for an electron acts only on one atom at a time; if the material composing a plate could be compressed into smaller volume there would be no change in the returned radiation.

CHAPTER IX

THE LOSS OF ENERGY OF THE β RAY

LET us now turn our attention to the second of the main causes of the alteration of a β ray stream, the actual loss of energy in passing through matter apart from scattering. It is still a matter of experiment and debate as to how this loss of energy takes place. It may be gradual as in the case of the α particle, an analogy which we cannot neglect. Or there may be large losses of energy at single encounters, as is sometimes assumed to happen in the case of cathode rays, on the theory that such sudden alterations in speed originate pulses in the æther which constitute the X rays. Sometimes again it has been supposed that β particles can penetrate into atoms without being able to emerge again.

This latter idea received a good deal of support from the want of success of such experiments as were designed to measure the loss of velocity of the β particle consequent on passage through a screen; it was difficult to establish the existence of the loss, and it was concluded that there was none. The gradual wasting away of a stream of β particles was ascribed to their absorption, one by one, by the atoms through which they tried to pass. But the actual fact is now established. For example, W. Wilson compared the velocities of β rays before and after passing through aluminium sheets and found a diminution which was greater the slower the rays. Half a millimetre of aluminium reduced the speed from $2{\cdot}85 \times 10^{10}$ cm./sec. to $2{\cdot}80 \times 10^{10}$ cm./sec., and again from $2{\cdot}735 \times 10^{10}$ cm./sec. to $2{\cdot}60 \times 10^{10}$ (*Proc. Roy. Soc.*, lxxxiv, 1910, p. 141). Wilson was unable to determine whether his results better fitted the formula $E^2 = k(a - x)$, where E is the energy of the ray, x the thickness of matter traversed, and k and a are constants, or the formula $E = k'(a' - x)$; but he inclined towards the latter.

Becquerel gives some figures which also fit in better with the second formula (*C.R.* cxxx, p. 372, Feb., 1900). Having separated out the β rays by a magnetic field into pencils of different velocities, he determined the minimum velocity which would carry the β ray through aluminium screens of different thicknesses with enough intensity to act on a photographic plate behind the screens. He found that for thicknesses 0·010, 0·100 and 0·200 of aluminium the velocities were in the ratio of 431, 1192, and 1746 respectively. The latter figures are the products of the strength of the field and the radii of curvature of the paths of the particles. Squaring them, it appears that the energies are in the proportion 1, 7·7, and 16·5. The fourth powers of the velocities are as 1 to 59 to 270: and the thicknesses of aluminium screen are much more nearly in proportion to the squares of the velocities. No account is taken of the effects of scattering in these experiments, but there is no large amount of it in aluminium.

There is a natural difference between the interpretations of the two different laws. If the loss of energy of the β particle is due to the imparting of energy to electrons which it passes by, then the fourth power law should be followed; for those β particles which go straight through a screen can have experienced only a number of very little losses of energy from the actions of the electrons in the atoms, no one of which has been sufficient to deflect it seriously. The momentum given to an electron by the β ray is proportional to the time the latter takes to pass by and therefore inversely to the velocity : and the energy given to the electron by the β particle is inversely proportional to the square of the velocity. Hence if E is the energy of the β particle, $\delta E = -k\delta x/E$ where k is a constant, and $E^2 = k(a-x)$ (Dunedin Address, 1904). The other formula, $E = k'(a'-x)$, implies that energy is somehow spent uniformly along the track, as is nearly the case with the a particle until its speed becomes small. Many investigations have also been made of the loss of speed of cathode rays in similar circumstances. Of these we may take as an example the recent work of Whiddington (*Proc. Camb. Phil. Soc.*, July, 1911), who finds that with aluminium screens there is good agreement between experiment and the fourth power formula

$$(v_0^4 - v_x^4) = ax,$$

where x is the thickness of matter penetrated. For aluminium Whiddington finds $a = 7{\cdot}32 \times 10^{42}$. This formula will take in satisfactorily the measurements of Wilson on the slower β rays, but not the rest.

The analogy of the a particle should certainly lead us to expect that the loss of energy takes place regularly along the track of the β particle, with greatly accelerating rapidity as the speed dies down. Remembering that the chances of deflection increase very rapidly with the speed, and that even in the case of the a particle they have a marked effect upon the end of its passage through matter, so that they must be far more effective in the case of the lighter β particle, we can form some idea of the complete path of the β ray. At first it moves fairly straight - with little deflection, it spends little energy, and produces little ionisation; but later on the deflections become more violent and frequent and the rate of expenditure of energy grows at the same time; and once begun the increases accelerate so much that the particle may seem to come to a sudden stop. It is simply an exaggerated case of a particle motion, and even taking in all the scattering the β particle has something like a definite range, as Becquerel's experiments show well (Fig. 36).

It does not seem that the act of deflection causes much loss of energy; we have already seen that the scattered rays have a velocity not far short of the primary. We may also draw this conclusion from a study of Madsen's curves (Fig. 39). If the ordinates of D and F, referring to gold, be added together, it will be seen that the new curve at first falls very slowly, if at all. When the screen weighs as much as 0·008 gram per sq. cm., the sum of the transmitted and returned radiations is still equal to the original stream; at the most, the loss on the scale adopted is 5 while the measure of the returned radiation is 37. In the case of aluminium the same conclusion can be drawn, but it only holds for a shorter piece of the curve.

The information we desire as to this gradual loss of energy will be capable of expression, when we have got it, in the form of a curve, in which the abscissæ will represent length of path in any given material and the ordinates energy which the particle must initially possess if it is to complete that path. The path here considered will not necessarily be straight—it may be very much broken up into zig-zags—but its actual length when straightened out is the quantity to be considered. There will be such a curve for each different substance which will express completely the rate at which electrons of all speeds will lose energy in that substance. Experiments such as Wilson's and Whiddington's give information as to different parts of the aluminium curve.

I have recently made some measurements of the relative lengths of the track of a given β ray in various substances by a method based on a quite different principle—it deals with the whole track, rectified in the way just described, and avoids the effects of scattering. It is based on some of the laws of the γ rays which we may anticipate sufficiently for the present purpose. These laws are (*a*) that γ rays of given quality give rise to β rays of a definite initial speed no matter in what material the β rays are produced, (*b*) that the production of β rays is proportional to the absorption of γ rays by the substance through which they pass (*Phil. Mag.*, Sept., 1910).

Suppose a block of any material to be crossed by a stream of γ rays. In every gram of the material a certain production of β rays takes place and these rays start their courses through the material, deflected at various encounters with the atoms they meet and also losing energy as they go. If in any material the loss of energy is rapid, the track will be short. By the track is meant, not the distance the particle succeeds in reaching from the point where it started, but the amount of matter penetrated.

Supposing the track were straightened out. A strict definition may be made in this way. Let the rectified track of the β particle be the axis of a cylinder having a cross-section of area a, and let the weight of the cylinder be w, then the amount of matter traversed, or the whole track, is w/a: we may call this the range of the β particle and denote it by d.

Consider a unit of volume in the material through which the γ rays are passing. and suppose the γ rays to be of equal intensity at all places in the material: if this condition is not quite realised in experiment, we can easily allow for the defect. The sum total of all the portions of the paths of β particles which are completed within the unit of volume in each second is proportional to two things, (*a*) the number of β rays originated in each unit of mass of the substance (this is nearly independent of the nature of the substance where the γ rays are very penetrating), (*b*) the range d. This would be true even if d were not actually the same for every ray but were only an average. This statement will perhaps be more obvious when it is considered that halving the density would halve the number of β rays springing up in each unit of volume, but double the range of each so that the number crossing each unit of volume in a second would be the same as before. In other words, the whole length of the tracks completed in a unit of volume depends only on the nature of the substance and not on its density or uniformity; it is quite defined by *Ikd*, where *I* is the strength of the γ rays, *k* the coefficient of absorption of the γ rays per unit mass of substance crossed by the rays, and *d* the range.

The important thing to observe is that the track need not be straight; the β particle may make any number of changes of direction during its whole range, but so long as *d* is measured as defined the quantity *kd* is unaffected by the crookedness. The experiment takes account of the loss of energy only and is independent of the scattering.

It may be of some service to give an analogy. If *k* points were taken at random in each square centimetre on a sheet of paper and a line of length *d* were drawn from each point, then the quantity of ink used and the quantity of ink on each square centimetre would be just the same, on the average, whether the lines were straight or curved or made up of any number of short pieces so as to be zig-zag in form.

The object of the experiment is to obtain a relative measure of the quantity *Ikd* in different substances. *I* being kept constant, and *k* being nearly independent of the nature of the substance, the result is a relative measure of *d*.

Suppose a cavity to be made in the substance. This makes no difference whatever in the value of *kd* anywhere within the boundary of the substance including the cavity once filled by the substance. This follows from the fact that every β particle has to cross a weight *d* of the substance; crossing the cavity does not count in the total path because there is no material in the cavity for the β particles to cross. The only source of experimental inaccuracy is due to the difficulty of making I the same in all parts of the substance that border on the cavity. But the difficulty is not serious. We may say, therefore, that just as many β rays would cross each unit volume of the cavity as would cross were the cavity filled with substance like the rest of the body or any other substance having the same kd. The shape of the cavity is immaterial. We may in fact take it to be the inside space of an ionisation vessel, provided only that the walls are too thick to be penetrated by the β rays.

It is curious but not uninstructive to consider that if we had a substance with no k, but with the power of reflecting every β ray that fell upon it, and made a closed vessel of the substance and shot γ rays across it, we should then get the following results. If a vacuum existed in the vessel, kd would be zero; if a single atom of any substance were placed in the vessel the value of kd throughout the vessel would in time become the full value for that substance and would not be increased if the one atom were added to by the addition of any number of the same kind. If atoms of other kinds were inserted there would be a compromise, the 'density' of the β rays then becoming $\Sigma k/\Sigma(1/d)$.

To go back to the line of argument, the introduction of air into the cavity in the substance under investigation makes little difference to the density of the β rays within it unless kd for the substance differs considerably from the kd of air, and there is so much air that an appreciable fraction of β ray energy is used up when a stream of such rays tries to cross the cavity. Hence the cavity must not be too big or the pressure inside too great. If there is any doubt about it in a given case, it can be tested by varying the pressure of the air; if the relation between pressure and ionisation requires a curved line to represent it, the initial portion of the curve must be used for which the pressure is small. This precaution is usually unnecessary and we may take the ionisation in the air of the cavity to be proportional to kd, and of course to the pressure.

If, therefore, we make a number of ionisation vessels of different materials but of the same form and cause γ rays to cross them, the amount of ionisation inside, other things being equal, is a measure of the value of kd for that substance, and therefore of the range in that substance of the β ray of a certain definite initial speed. The experiment is conveniently carried out by making a thick lead ionisation chamber—the thick lead cuts out all but the penetrating rays—and inserting linings of different materials. The γ rays must be kept at the same strength inside each lining, or if not, any differences must be allowed for. The following table shows results obtained in this way. The first column gives the material of the lining, and the second its thickness, which was enough to give the true value of kd except possibly in the cases of cardboard and aluminium. The third column gives the results obtained when the γ rays had to pass through little more than the lead wall of the chamber, which was 0·47 cm. thick, and the fourth the results when the rays had also to pass through a lead screen 1·1 cm. thick. The figures are corrected for differences in volume and for differences in the strength of the γ rays due to absorption in the linings.

TABLE XV.—*Relative Ranges of β Rays.*

I. Substance.	II. Thickness of lining.	III. Ra unscreened.	IV. Ra screened.
Lead	...	100	100
Tin	0·16	58	68
Zinc	0·21	47	55
Iron	0·155	45	54
Aluminium	0·21	40	49
Card	0·24	39	46

The height of the chamber, which was cylindrical in form, was 15 cm. and the diameter 9 cm. The radium was placed on the axis of the cylinder, 10 cm. away from one end.

The differences in the figures in the last two columns are really due to a change in the relative value for lead only. The rays have been so hardened by passing through the extra cm. of lead that the absorption coefficient (k) of the lead lining has fallen to the same value as that of the other metals. In the first case, there is a special production of softer β rays by the lead which does not take place in the second.

The figures in the last column are relative measures of the range of the β particle in the different substances opposite to which they stand. It is not really a matter of surprise that they decrease with the atomic weight, for so do the ranges of the α rays diminish in the same way when measured in terms of the weight of matter penetrated. The stopping power of the atom for α rays is nearly proportional to the square root of the atomic weight; for the β particle it is rather proportional to the cube root. In the case of the α particle we have seen how the stopping power depends a little upon the speed; in the case of the β the corresponding question has been little considered.

The third event of importance in the history of the β particle is its occasional disappearance in favour of an X or γ ray. The production of X rays by the cathode stream is the most striking example of it, but not much is known about the effect quantitatively. As regards the production of γ rays from β rays, it is only lately that an example has been found by Gray, who has shown (*Proc. Roy. Soc.*, 85, 1911, 131) that the β rays of RaE produce γ rays in appreciable quantity. It will be most convenient to postpone the consideration of these effects until we have to do with the X and γ ray phenomena generally.

CHAPTER X

THE GENERAL CASE OF 'ABSORPTION' OF THE β RAY

SUPPOSING then we knew all the facts relating to scattering, to loss of energy, and to conversion into X or γ rays, it would still be a most complicated problem to calculate the quantities of β radiation transmitted and returned by a given screen: this we have already seen. It is worth while to add that corresponding complexities occur in the interpretation of the currents found when a stream of β rays enters an ionisation chamber. The chamber is generally large enough to take in electrons moving in various directions and most probably with various speeds. The walls of which it is composed—no matter what the substance of which they are made—return a fraction of the energy that falls upon them, a fraction which varies with any change in the quality of the rays; and the ionisation of the gas in the chamber depends on the velocity and length of path within of every separate electron that enters it.

We cannot, on all these grounds, expect a simple law to cover the results of any particular experiment. If, for example, we place screens of varying thickness in the path of a pencil of β rays and find in each case the ionisation in a chamber on the far side of the screen we can plot an absorption curve which may approach the exponential form or may depart from it more or less. But even if an exponential curve does appear, there is usually no simple physical interpretation. It is convenient to assume an exponential law at times because it is the curve which, having an ordinate declining at a constantly decreasing rate, has also an extremely simple form of mathematical expression. But when accuracy and appropriateness of representation are required, the exponential curve and the coefficient of absorption derived from it may be as misleading as in the case of the α rays.

If the calculation of the effects of thick screens from the fundamental properties of the β ray is a complicated problem, still more difficult is the converse, the determination of the fundamental properties from measurements with thick screens. It is only by making large assumptions or rough approximations that there is any hope of a solution at all. Nevertheless, very interesting and important investigations have been carried out on these lines: on which, in fact, most workers have proceeded. Let us consider examples.

H. W. Schmidt has forced the problem within the reach of calculation by supposing the scattered rays to go only forwards or backwards in the line of motion of the primary ray and to undergo no change of velocity at scattering, by leaving out any gradual change in the properties of the β particles due to passage through matter, and lastly by supposing β particles to disappear while going at full speed so that their number gradually diminishes. (*Jahrbuch der Rad. u. Elek.*, Bd. V., Heft 4). A stream of β rays while passing through a layer δx of absorbing screen is supposed to lose two fractions of its energy, (1) a fraction $\beta\delta x$ which is turned right back through 180°, (2) a fraction $a\delta x$ which disappears altogether. This leads to values for the fractions of energy returned (ρ) and transmitted (δ) by a plate of thickness x

$$\rho = \frac{p(1-e^{-2\mu x})}{1-p^2e^{-2\mu x}} \quad ; \quad \delta = \frac{e^{-\mu x}}{1-p^2e^{-2\mu x}}(1-p^2),$$

where $a=\mu(1-p)/(1+p)$ and $\beta=2\mu p/(1-p^2)$. Where x is very small, these become $\rho=\beta x$, and $\delta=e^{-\mu x}$.

Schmidt has applied these formulæ to the results of a large number of experiments made with the β rays of uranium, and has found them to represent the observed facts fairly well, especially the variation of the transmitted rays with the thickness of the screen, for all the metals used as the material of the screen. The only serious discrepancies occur in the values of the returned radiations for thin screens of small atomic weight, and it is just here that the roughness of the approximation might be expected to tell. The returned radiation from aluminium is certainly not the same in quality as the incident; it is softer and produces more ionisation per cm. than primary rays of the same total energy (see p. 79). Schmidt's experimental values were nearly twice as large for this metal as his theory seemed to require.

Schmidt found that a/D varied very nearly as the cube root of the atomic weight, and β/D was nearly proportional to the weight, D being the density of the substance. Now Schmidt's constant a is derived from the definition that the β ray pencil loses a fraction $a\delta x$ of its energy in going through a distance, δx, or a mass $D\delta x$. If we change over the physical conception from that of a stream of rays diminishing in number but not in speed to that of rays diminishing in speed but not in number a/D is an inverse measure of what I have called the range of the β particle (p. 95). Consequently a/D should vary inversely as "*kd*": and the following table shows that this is the case:—

TABLE XVI.

Substance.	a/D.	kd.	$a/D \times kd$.
Lead	1·69	100	169
Tin	2·14 (2·40)	68	145 (163)
Zinc	3·00	55	165
Iron	3·08	54	166
Aluminium	3·26	49	160
Card	3·32*	46	153

* Calculated as for carbon from later figure given by Schmidt.

The uniformity of the figures in the last column is only broken seriously by tin. Strange to say, the value 2·14 which Schmidt gives for tin is quite out of line with the values he gives for all the other metals; if these are plotted and a value for tin obtained from the curve, we get 2·40, which leads to a value 163 in the last column. Schmidt's values of a/D for the β rays of actinium do not show this irregularity in the case of tin.

Thus the values of a found by Schmidt are in general agreement with what we should expect from a measurement of the range or of the loss of energy in passing through matter, apart from scattering. We shall see presently (p. 104) that the values of β give a satisfactory account of the scattering apart from the loss of range.

McClelland has also made an extensive series of researches on the same subject. His method of attacking the problem is similar to Schmidt's in mathematical form, but differs essentially in principle; in particular, he emphasises the existence of a true secondary radiation to be distinguished from the scattered primary and having a different origin. I have already tried to show that the simpler view is sufficient, and I have discussed McClelland's interesting experiments in the *Philosophical Magazine*, Sept., 1910, p. 391.

An entirely different scheme of approximations has been devised by J. J. Thomson, and made the basis of experiment by Crowther (*Camb. Phil. Soc. Proc.*, xv., 5, p. 442). It is supposed that when β particles are directed normally against a plate, they are gradually turned away from their initial direction through an accumulation of small deflections. It is also assumed that all these deflections may be treated as equal to their mean, and that the aggregate deflection is only small. No account is taken of any change in the velocity, or of any diminution in the numbers of the particles; thus scattering is the only effect considered.

I have discussed this method in the paper just referred to (*Phil. Mag.*, Sept., 1910, p. 414). The assumptions involved seem to me to be open to much objection, and the experimental results to be more in agreement with the views set forth in the preceding pages. When the thickness of the absorbing screen is gradually increased from a small value, the first signs of scattering are due to single deflections, except possibly in the case of those particles the directions of which have been changed only slightly. None of the experimental results can be ascribed with certainty to such "compound scattering" as Thomson supposes. All the much deflected particles, the screen being thin, illustrate "single scattering"; and this leaves only a small region of slight deflection, in which experiment is very difficult and interpretation dubious.

Rutherford has advanced a very interesting view of the scattering process (*Phil. Mag.*, May, 1911). The atom is assumed by Thomson to consist of a sphere of uniform positive electricity, containing negative electrons at various points within it, the charges balancing. Rutherford points out that such an atom is wholly unable to deflect the α particle to the extent which experiment reveals; on the other hand, if the positive is collected at the centre of the electron system, the magnitude of the charge being $-ne$, where n is of the order of the atomic weight and e the charge of the electron, the atom so constituted will deflect the α particle to the observed amount for all angles. This also gives the right amount of scattering for the β particles, making it proportional to the atomic weight as Schmidt found (*loc. cit.*, p. 486). On this view, the deflections of the α particles and of the β particles for thin screens are the result of single and not compound scattering.

The experiments of Lenard also demonstrate this point. From Fig. 1, it is clear that the pencil of cathode rays is not steadily opened out, but that individuals successively leave the main stream.

CHAPTER XI

GENERAL PROPERTIES OF X AND γ RAYS

LIKE the α and β rays, the rays which Röntgen discovered are able to bring about ionisation, phosphorescence, and photographic action, and these properties enable us to detect their presence, to follow their motion, and generally to investigate their properties. They differ materially from the rays which carry electric charges, because they are not affected when they pass through electric or magnetic fields: they are therefore uncharged. They move in straight lines and cast sharp shadows of objects through which they cannot pass. They are gradually absorbed in passing through any form of matter, and there is very good reason to believe that a homogeneous pencil of rays is absorbed according to an exponential law. Consequently, the penetration of rays of a definite quality through a given substance is expressed by an absorption coefficient λ, having the meaning that there is a fractional loss $\lambda\, dx$ when the rays pass normally through a sheet of the substance weighing dx grams per sq. cm.

It is better to use mass coefficients rather than linear coefficients, because the mere geometrical arrangement of the atoms or molecules of the absorbing substance is of no consequence: the mass alone influences the absorption. The absorption coefficients vary very greatly; rays have been observed which are unable to penetrate more than a few centimetres into air at ordinary pressure and temperature, and others which are only half absorbed in twenty metres and more. Rays are sometimes classified by their power of penetrating aluminium, a method which has the usual disadvantage of the arbitrary standard, but is convenient and simple. There are remarkable anomalies among the absorption coefficients. Homogeneous rays of such a quality that their absorption coefficients in Al is 71·6 are rather less absorbed by Fe (67·2), while rays which penetrate Al somewhat more easily ($\lambda = 59{\cdot}1$) are highly absorbed by Fe ($\lambda = 314$); and other instances can be found from Barkla's table, which is given on p. 152. Consequently, rays which are heterogeneous are of little use to the investigator.

Absorption is entirely a volume effect: no surface action has ever been proved. The rays pass just as well through glass when it is powdered as when it is in the form of a solid sheet. Nothing is to be found of the nature of reflection or refraction. The independence of physical and chemical condition is carried still further, since the absorption coefficient of any compound can be calculated additively from the coefficients of the separate atoms which are included in the compound. The absorption of the X ray is a process in which the action and reaction between the ray and each atom is not influenced by the condition of the atom, or by any relations between the atom and its neighbours of the same or other molecules.

The absence of reflection and refraction give the rays more resemblance to the α and the β rays than to the radiation of light, and there is a still further similarity to the material rays in the way in which the X rays are irregularly scattered by every substance through which they pass, the effect being again atomic, and not molecular. And again, their passage through any substance shows no dependence upon any crystalline structure which the substance may possess; they do not exhibit the phenomenon of double refraction nor is there any evidence that they can show any diffraction effect; if they could, it would be hard to resist the conclusion that they consisted of some vibratory motion, and the experiment would be looked to for a determination of the wave length. Haga and Wind in 1901 described an effect which they had observed (*Phys. Zeit.*, ii, p. 292) and supposed to be due to diffraction of extraordinarily short waves, but it has been seriously questioned by Walter and Pohl (*Ann. d. Physik,* xxix, 2, p. 331, 1909), and was no doubt misinterpreted. The latter authors showed that if there is a wave length of the X rays it was less than $1{\cdot}2 \times 10^{-9}$ cm. in the case of the rays which they employed.

The rays can be polarised in a certain sense. Barkla has shown (*Phil. Trans.*, cciv, A, p. 467) that a pencil of rays may be in such a condition that it is more likely to be scattered in some directions than in others. It can possess, as a beam of light can possess, a transverse polarisation. For example, in a case given by Bassler (*Ann. d. Phys.* xxviii, 4, p. 880, 1909) a pencil of primary rays moved from the X ray bulb at right angles to the cathode rays and fell upon a block of paraffin. Measurements were made of the amounts scattered in various directions in a plane perpendicular to the direction of the X rays, and it was found that the amount scattered in that plane and parallel to the cathode rays was less than the amount scattered in the perpendicular direction in the same plane in the proportion 90 to 110.

At this stage it is convenient to take into consideration the γ rays also: for experiment is continually suggesting that X and γ rays are similar in kind though quantitatively their properties may differ considerably. The γ rays are not influenced by electric and magnetic fields: they are not reflected, nor do they show refraction, interference, diffraction or double refraction. They differ from X rays mainly through their extreme powers of penetration and in the manner in which they make their appearance.

Otherwise they resemble each other so closely that it is convenient sometimes to include both under the general heading of X rays. This assumption of their similarity in nature will be found to be justified as we proceed.

The æther pulse theory of the X and γ rays is based on the assumption that they consist of disturbances in the æther which spread away from places where electrons have received accelerations, positive or negative, of their motion. Stokes, Thomson, and other workers have attempted to mould the theory so that it will fit in with the properties outlined above.

There remains, however, one striking phenomenon in connection with these rays, which we have not yet considered, and which was not, as we now see, taken into proper account by the authors of the pulse theory. This is the emission of swiftly moving electrons from matter through which the X or γ rays are passing. A very brief consideration of this effect will show its importance.

The swiftly moving electrons may have so great a speed that they can be ranked with the β rays, the properties of which we have already considered; we find speeds of this high order in the case of the electrons owing their motion to the γ rays. Or, again, they may be slower, having initial speeds varying between 10^9 and 10^{10} cm. per second. The electrons ejected under the influence of X rays have velocities of this order, and are generally called cathode rays, since these are the usual velocities of the cathode rays in the vacuum tube. But whether they are possessed of the greater or the lesser speeds they are all able to ionise a gas, to act on a photographic plate, and to excite phosphorescence in suitable materials, and it will be convenient to call them all β rays irrespective of the order of their velocity, in the same way as we have decided to use the term X ray to cover the γ ray also, unless, of course, there is reason to make a distinction.

Now it is clear that if X rays produce β rays, and if the latter are able to cause ionisation, phosphorescence, and photographic action, then there is no evidence, *a priori*, that the X rays themselves possess this power. It may be that the effects which we find in the presence of X rays are not due to the rays themselves directly, but to the β rays which they produce. This doubt would in any case prompt us to begin a thorough investigation of X ray phenomena by considering the circumstances in which the one kind of ray produces the other, and by examining the relations between the qualities and properties of the X rays on the one hand and on the other the speed of the β rays and the nature of the matter in which they arise. This we will now do. It will be seen later that we could have followed no other course, for the doubt turns out to be quite justified.

The evidence which we shall consider goes to how that the X ray has no direct action whatever except the one of causing the β ray. Only in this way does it reveal its existence. Up to the point when the β ray is produced the X ray spends no energy and causes no observable effect. After that point has been passed, the effects are those of the β ray, and as such have been considered already. There is left for the main object of our consideration the moment of the production of the β ray. This is an anticipation, and I have inserted it with the object of making it easier to keep in mind the arguments that follow. For the same reason, it may be well to add that this method of procedure does not lead immediately to the æther pulse theory. This theory, as I have already said, was formed at an earlier stage with the object of explaining the first found properties of the rays—their penetration, their lack of reflection, refraction, and so on. But it does not give a ready account of the circumstances peculiar to the production of β rays: and if it is on these that attention should be concentrated, if there is really little else to consider as the direct result of X ray action, then a theory which does not take them seriously into account is not immediately helpful.

CHAPTER XII

THE PRODUCTION OF THE SECONDARY β RAY BY THE X RAY

CURIE and SAGNAC showed in 1900 (*Comptes rendus*, cxxx, p. 1013) that some of the secondary radiations emerging from a plate struck by Röntgen rays were negatively electrified. Dorn was the first to attempt a measurement of the velocities of the negative rays (*Archives Néerlandaises*, v, p. 595, 1900). He used the method of deflection by a magnet, and found velocities varying from $1 \cdot 8 \times 10^9$ cm./sec. to $8 \cdot 5 \times 10^9$ cm./sec. Since the immediate result of his experiments was the determination of the compound quantity mv/e, where the symbols denote the mass, velocity, and charge on the ray, it was necessary to assume a value for e/m, in order to be able to calculate the velocity. Dorn assumed that the rays were electrons in motion and therefore used the value for e/m which had recently been found for the electron. There can be no doubt that these rays really are what Dorn supposed them to be, although the charge of the carrier has never been actually measured; for the general properties of the rays are exactly those of cathode rays, just as the general properties of the secondary β rays due to γ rays are those of primary β rays; the powers of penetration into various materials are in complete agreement with the calculated values based on the assumption that they are electrons in both cases. Dorn used an X ray bulb of average hardness; but he gives no exact definition of its state, nor does he record measurements of the effect of varying the state. Since it is now known that the velocity of the β ray depends materially on the quality of the primary X ray, it is better to push on to the consideration of later work.

In 1907 Innes attempted a more thorough investigation, including the consequences of varying (*a*) the hardness of the X ray bulb, (*b*) the nature of the substance from which the secondary β rays emerged, (*c*) the distance between the bulb and the metal. His results included some of great importance, which we must now consider (*Proc. Roy. Soc.*, lxxix, p. 442).

The hardness of the X ray bulb was measured in terms of the equivalent spark-gap. The bulb was placed at various distances from the metal plate upon which the X rays were allowed to fall. To reach the plate they had to pass through a thin aluminium sheet (0·1 mm. thick) which covered an opening in an evacuated chamber containing the metal. It was necessary to exhaust this chamber of air; for the β rays due to Röntgen rays can only penetrate a millimetre or two into air at ordinary pressures, and it is only in a good vacuum that they move without appreciable scattering over such distances that their paths can be readily determined. The β rays passed in succession through two slits cut in lead plates and fell upon a photographic plate, the whole of this apparatus being enclosed within the chamber.

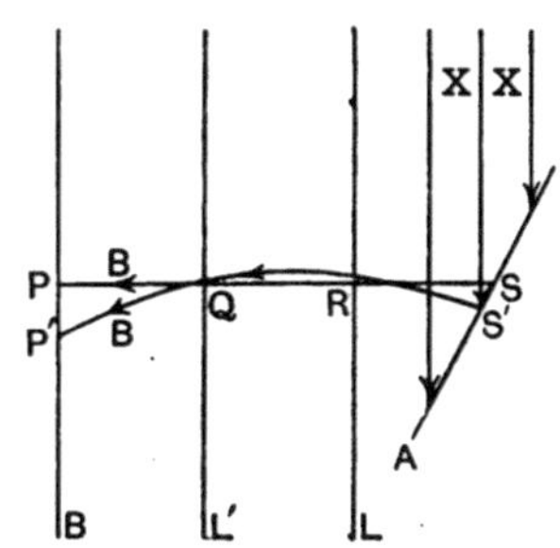

FIG. 42.

The arrangement may be represented diagrammatically as in Fig. 42, where B is the photographic plate, LL' the lead plates with their slits, and A the metal plate which was the source of the β rays when X rays fell upon it. An impression was made at P by the rays which had proceeded in the straight line *SRQP*. When a magnetic field was applied, generated by the current in a pair of Helmholtz coils, the rays which struck the plate had pursued a circular path, *S'RQP'*. From the distance between *P* and *P'*, the dimensions of the apparatus, and the strength of the field the velocity of the rays could be calculated. The value of e/m was assumed to be $1 \cdot 7 \times 10^7$.

One of the most important results of these experiments is the proof that the speed of the secondary β ray does not depend on the distance between the bulb and the plate; it is independent of the intensity of the *X* radiation. Innes made further tests of this deduction by varying the current passing through the *X* ray bulb, and again by varying the frequency of the interruptions of the current in the primary of the induction coil. In no case was there any effect upon the velocity. The following figures will serve as an illustration of effects of varying the distance between bulb and plate. The first column gives the nature of the plate :—

TABLE XVII.—*Showing speed of secondary cathode ray to be independent of intensity of primary X ray.*

Substance.	Distance.	Velocity × 10^{-9}.
Lead	6·4	7·6 to 6·4
,,	19·3	7·6 to 6·3
,,	50·0	7·8 to 6·2
Silver	6·0	7·2 to 6·0
,,	20·0	7·3 to 6·0

It will be seen that very little change in the velocity is to be observed, although in the case of lead the intensity must have been about sixty times as great in the third experiment as in the first.

Other interesting relations are to be found among the experimental results which Innes obtained. An increase in the spark-gap produced an increase in the speed of the fastest of the secondary β rays. This is clearly shown in Table XVIII.

TABLE XVIII.—*Variation of Speed of Secondary Cathode Ray with Quality of X Ray.*

Substance.	Spark-gap.	Velocity × 10^{-9}.
Lead	5·1	7·8 to 6·3
,,	16·0	8·3 to 6·4
Silver	3·9	7·2 to 6·0
,,	19·0	8·0 to 6·1
Platinum	3·2	7·4 to 6·1
,,	14·0	8·0 to 6·5
Gold	3·4	7·5 to 6·1
,,	15·0	8·1 to 6·2

We may express this result by saying that when the cathode rays in the X ray bulb move faster, so also do the secondary β rays.

The nature of the plate also appeared to have some influence, for when zinc was used, the velocity limits were narrowed down to $6{\cdot}4 \times 10^9$ as the upper, and $6{\cdot}0 \times 10^9$ as the lower of the two. It will be observed that the other four metals all behaved alike. Later evidence confirms the variation of speed with the hardness of the tube, but shows that the nature of the metal has only an indirect effect.

The range over which Innes was able to vary the quality of the X rays was very small in comparison with what is now possible, and it is better to defer any further consideration of these questions until we have examined the results of later investigations.

We may next consider some experiments made by Madsen and myself (*Trans. Roy. Soc.* of *South Australia*, Jan. and May, 1908, or *Phil. Mag.*, May and Dec., 1908). Our purpose was to investigate the production of β rays from γ rays in order to determine in the first place the extent to which the speed of the β ray depended upon the quality of the γ ray, in the second, the extent to which the speed depended upon the nature of the atom, and in the third, any relations that might exist between the initial direction of movement of the β ray and the direction of the γ ray to which it was due. With regard to the last of these questions, it is easy to arrange experiments which show that there is much to be examined. Such experiments are given in the earlier of the two papers referred to, and they may now be described briefly.

Let us imagine a stream of γ rays passing normally through a very thin absorbing sheet; for example, through a millimetre of aluminium. Secondary β rays emerge from the sheet on each of its two sides, and it becomes a matter of interest to inquire whether the two quantities are equal. Let us describe as "incidence rays" those which appear on the side where the γ rays enter the sheet, and "emergence rays" those which appear on the other. It is not easy to measure the energy of either radiation with any accuracy, to compare it, that is to say, with some proper standard such as the γ ray energy which is absorbed in crossing the ionisation chamber. We may confine ourselves at first to the devising of a simple experiment which will show that the emergence and the incidence radiations, due to a stream of γ rays passing through a sheet, are far from being equal to each other in all cases. Consider an ionisation chamber shaped as in Fig. 43. The two ends are closed by plates of which A and A' are made of the same substance and are of the same thickness. The two plates, B and B', are made of a different substance, but are similar to each other. Let a pencil of γ rays pass along the axis of the chamber, which is represented by a dotted line. The ionisation current within the chamber is measured as usual by an electrode connected to an electroscope.

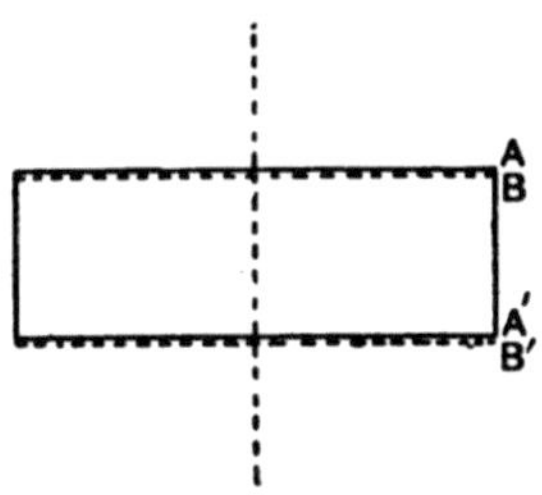

FIG. 43.

If the β radiations on each side of a sheet were equal to each other, then the pencil of γ rays would give rise to the same amount of secondary β radiation in the chamber, no matter which way it went, upwards or downwards. In each case there is β radiation from one side of each metal, caused by practically the same pencil of γ rays. We can say the same pencil; for if the rays are going downwards, let us suppose, such β rays as come from B have their origin in a layer of B which is so thin that very little γ ray energy is absorbed in crossing it, and the same may be said of the air in the chamber and of the thin sheet on the other side in which the β radiation of A' takes its rise. But when the experiment is made, there is a very large difference in the current in the two cases, and it is clear that in the case of at least one of the substances used the β radiation on one side is greater than on the other.

In one of the experiments quoted in the paper (see Fig. 44) the chamber was of cylindrical form, 3 inches high and 10 inches in diameter. The substances used were aluminium and lead. The thickness of each plate was a little less than 2 mm. The radium was placed at the narrow end of a conical hole in a lead block and so arranged that the rays could pass along the axis of the chamber, approximately. They entered through plates of Pb and Al, A and B, which could be transposed so as to alter the order in which the rays went through them; and they fell upon plates A' and B', which could also be inverted. Inversions of the plates were simpler to effect in practice than a change in the position of the radium. With this arrangement inversion of the top plates A and B made a difference in favour of Al of about 1%, *i.e.*, the current was slightly larger when the Al plate was next the chamber. The emergence β rays were therefore somewhat greater for Al than for Pb. It is to be observed that the

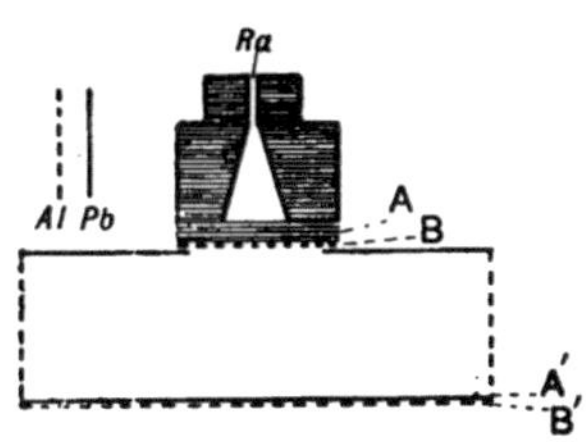

FIG. 44.

inversion of the plates A and B did not alter the quantity of γ rays that were active; the rays had to pass in each case through the same materials, and undergo the same absorption before entering the chamber.

When the plates A' and B' were transposed, the current in the chamber was 44 per cent. greater with the lead uppermost, which showed that the incidence β rays from Pb were much larger than the incidence rays from Al. We conclude that the emergence and incidence radiations are unequal, either in the case of Al, or of Pb, or of both metals.

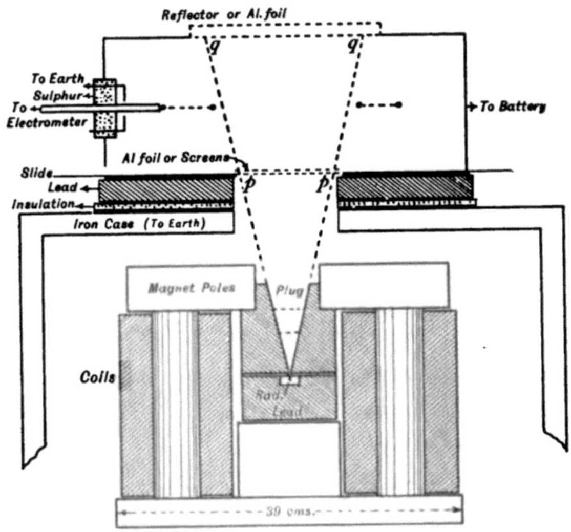

FIG. 45.

The apparatus employed by Madsen and myself in the more complete investigation is shown in Fig. 45. It is taken from a drawing in the original paper (*Trans. Roy. Soc. of South Australia*, xxxii, May, 1908, and *Phil. Mag.*, Dec., 1908). The radium was placed at the bottom of a conical hole made in a massive lead block. Plugs of various materials and thicknesses were turned to fit the hole, and could be used to investigate the effects of partially absorbing the γ rays by different substances.

The γ rays passed up between the poles of a powerful field magnet, which turned aside all β rays from the stream. It is to be remembered that a pencil of γ rays always contains β rays arising from the action of the γ rays on the last material through which they have passed, or the next which they are about to enter, and it is impossible to remove these β rays by a screen, since the interposition of material will only be the occasion of fresh secondary radiation which will take the place of that which it has intercepted. Nothing but a strong magnetic field and a special arrangement of the apparatus can even approximately remove the β rays from a space through which the γ rays are passing. It is to be remembered also that the air itself absorbs γ rays with a consequent production of β rays, so that on this account alone it is impossible to provide a space where γ rays are present without β rays.

A thick iron case was placed round the magnet and the γ rays passed up through a circular opening cut in the top of it. The iron prevented the stray lines of magnetic force from entering the ionisation chamber, and there distorting and altering the paths of the β rays inside it. On top of the iron case was a thick lead screen, intended to assist in preventing stray γ radiation from entering the chamber except through the opening provided. Above this was the ionisation chamber made of thin zinc. Sufficient openings were left at the top (qq) and at the bottom (pp) of the chamber to allow the rays to pass through without touching the zinc, and the openings could be closed by suitable screens of various substances. When the screens were made of the thinnest Al leaf supported on fine Al wires, the secondary β rays in the chamber were reduced to a minimum, though they were far from being negligible.

In the case of each substance investigated, measurements were made of the ionisation current under three different arrangements, namely :—

(*a*) when the screens *pp* and *qq* were of the thinnest Al foil;

(*b*) when *pp* consisted of a plate of the given substance just thick enough to give the full amount of emergence β rays;

(*c*) when a thick plate of the same substance composed the screen *qq*.

The differences between *b* and *a*, and between *c* and *a* were taken as measures of the emergence and the incidence β radiations respectively. The results are given in the following table, in which the two sets refer to soft and hard rays respectively. In the former case the γ rays were unscreened except by the walls of the capsule containing the radium, which were of light materials a millimetre or so in thickness. The hard rays were those left after passing through a thick lead plug inserted in the conical hole and were mainly composed of the more penetrating constituents of the original bundle of γ rays. The units are arbitrary.

TABLE XIX.—*Comparison of Emergence and Incidence β Rays.*

	Soft γ rays.		Hard γ rays.	
	Incidence.	Emergence.	Incidence.	Emergence.
C	170	2,280	58	1,150
Al	280	1,810	120	795
S	340	1,575	154	685
Fe	487	1,350	163	560
Cu	558	...	202	523
Zn	618	1,160	224	485
Sn	1,051	1,170	333	303
Pb	1,723	2,001	497	470

These figures show the existence of large dissymmetries in various cases which are far beyond the reach of experimental errors, though the latter are considerable. There are difficulties of interpretation, for it is evident that $b-a$ (the emergence radiation) is generally underestimated, since the screen *pp* stops a certain amount of

β radiation which is made in the lead just below it and in the air also, and which is reckoned in *a* but not in *b*. The screen also cuts down the γ rays themselves to some extent. And again the screen *qq* not only gives rise to the true incidence radiation but also turns back some β rays striking it from below, and this is most serious when the substance of the screen is of high atomic weight. It is difficult to allow for all these sources of error. They are almost necessary consequences of the very penetrating nature of the γ radiation which cannot be altogether limited to the pencil used in the investigation. It is difficult to purify the pencil from β rays and also to keep the ionisation chamber from being affected by secondary radiation springing up in surrounding objects where they are struck by escaped γ rays. The whole experiment is worth repeating on a still more massive scale, when the errors might be sensibly reduced.

From these results we can draw the general conclusion that the emergence radiation is greater than the incidence, particularly in the case of the substances of small atomic weight. The increase of incidence radiation with atomic weight is in agreement with earlier proofs of the same principle (Eve, *Phil. Mag.*, Dec. 1904; McClelland, *Trans. Roy. Dub. Soc.*, March, 1905). The relative amounts of incidence β rays from different substances when exposed to β and to γ rays are not very different, as may be seen from the table given by Eve (*loc. cit.*):—

TABLE XX.—*Comparison of Secondary β Rays due to Primary β and Primary γ Rays.*

Radiator.	β and γ rays.	γ rays.
Pb	100	100
Cu	57	61
Brass	58	59
Zn	57	...
Al	30	30
Glass	31	35
Paraffin	12	20

Madsen and I next proceeded to measure the quality of the secondary β rays in terms of their penetrating powers. When the thickness of the screen *pp* is gradually increased from the smallest value possible, the β radiation increases rapidly at first but reaches a maximum when the screen is so thick that β rays from its lowest stratum are unable to make their way through it, or indeed a little before this point if account is taken of the absorption of the primary γ rays. The results of experiments of this kind in which Pb and Al screens have been used are shown in Figs. 46 and 47. From these curves we can obtain an approximate

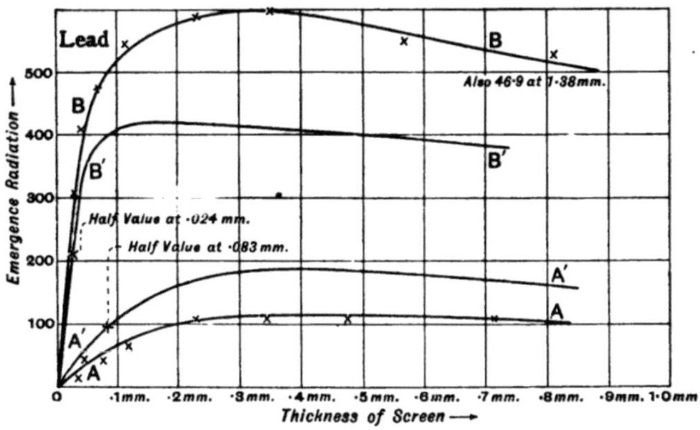

FIG. 46.—*Absorption Curve of Secondary β Rays in Lead.*

value of the penetrating power of the secondary β rays; for if the thickness of the screen be x and the coefficient of absorption of the β rays in the material of the screen be λ, the energy of the β radiation will contain a factor $\int_0^x e^{-\lambda x}dx$ or $(1-e^{-\lambda x})/\lambda$. This is obtained on the assumptions that the screen is not thick enough to absorb the γ rays appreciably, and that the β ray absorption follows an exponential law. When the screen is of that thickness for which the β radiation is half its full value $e^{-\lambda x}=1/2$ and $\lambda x=0{\cdot}7$.

In Fig. 46 the curve marked *A* shows how the emergence β radiation from Pb increases with the thickness of the

Pb screen at *pp*, when the γ rays have been sifted of their less penetrating constituents by a lead plug 1·6 cm. thick placed in the conical opening above the radium. The rays which are left after penetrating this amount of lead have a mass absorption coefficient in lead of about 0·037 (McClelland) or 0·041 (Soddy and Russell), the initial coefficient of absorption being about fifty per cent. greater. The β radiation is shown by curve *A* to rise to half its full value when the thickness of the lead screen is 0·083

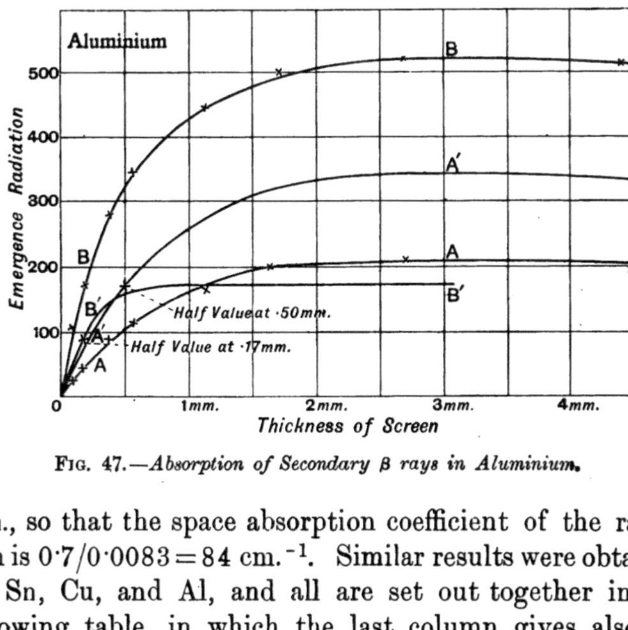

FIG. 47.—*Absorption of Secondary β rays in Aluminium.*

mm., so that the space absorption coefficient of the radiation is $0·7/0·0083 = 84$ cm.$^{-1}$. Similar results were obtained for Sn, Cu, and Al, and all are set out together in the following table, in which the last column gives also the values of the absorption coefficients of the primary β rays of radium in these substances, as given by McClelland and Hackett (*Trans. Roy. Dub. Soc.*, March 22, 1907).

TABLE XXI.—*Showing quality of Secondary β Rays due to Primary γ Rays to be the same as that of the Primary β Rays.*

I. Substance.	II. Thickness of screen giving half value in mm.	III. λ calculated from II.	IV. λ for β rays.
Lead	0·083	84	93
Tin	0·141	50	52
Copper	0·137	51	55
Aluminium	0·50	14	14

Since the figures in the last two columns agree so closely, we conclude that the secondary β rays, no matter in what substance they arise, have the same penetration as the primary β rays. It is true that hard γ rays were used in these experiments, while the values of the absorption coefficients of the β rays refer to rays of the usual quality belonging to RaC. The allowance to be made for this difference is uncertain; but it cannot be great, for the value of λ varies rapidly with the speed of the β particle, and there is room for considerable alteration of the values of λ in the third column of the table without thereby making much change in the corresponding velocities.

The curves marked B in Figs. 46, 47, exhibit the results of measurements of the emergence radiations when the lead plug had been removed, and the γ rays were therefore of a less penetrating character. Comparing them with the A curves, we have an opportunity of judging the effect of varying the quality of the γ rays. We see that the B curves rise much more quickly to their maxima, and indeed in the case of lead the total ionisation in the chamber begins actually to decline when the lead screen is only half a millimetre in thickness. This peculiar behaviour of the lead is due to the fact that the softer constituents of the γ rays are absorbed with especial rapidity by substances of large atomic weight. A lead screen "hardens" a heterogeneous beam of γ rays far more than a screen of aluminium. This is in agreement with the rapid decline of the absorption coefficient of the γ rays of radium as they pass through greater and greater thicknesses of lead: a decline which is not manifested with corresponding clearness in the case of the lighter atomic weights. As an example, we may take the following figures due to McClelland (*Phil. Mag.*, July, 1904).

TABLE XXII.—*Mass Absorption Coefficients of γ Rays.*

Substance.	Thickness of screen.			
	0·25 cm.	0·05 cm.	1·0 cm.	1·5 cm.
Pb	0·056	0·049	0·042	0·037
Al	0·038	0·038	0·038	0·038

The heterogeneous γ rays causing the effects which are represented by the *B* curves contain both hard and soft rays, though there is no clear line of distinction between them. It is possible to separate out the effects of the soft rays in the following way.

The effects of the hard radiation which has passed through the Pb plug is shown by the curve *A*. If the plug had not been there, the effects of this radiation would have been greater by about two-thirds, a result easily calculated. The curve *A′* is obtained by increasing the ordinates of *A* by two-thirds of their values, and the ordinates of *A′* are subtracted from those of *B*. We thus obtain *B′*, a curve which may be taken to represent roughly the quality of the β radiation due to the less penetrating γ rays. The radiation now rises to half value when the lead screen at *pp* is only 0·024 mm. thick, which is not much more than a quarter of the corresponding value for the hard rays. The more penetrating constituents of the γ rays of radium give rise to secondary β rays which are four times as penetrating as those due to certain of the less penetrating portions of the same radiation.

The results of this investigation upon the relation between the secondary β ray and the γ ray to which it is due may therefore be summed up in the following statements:—

(*a*) The velocity of the β ray depends on the quality of the γ ray, increasing with the penetrating power of the latter.

(*b*) The velocity is independent of the nature of the atom in which the β ray arises.

(*c*) The emergence β radiation is generally greater than the incidence, particularly in the case of the light atoms.

It is possible to put the last statement in a different form, which throws some light on the manner in which the β ray begins its motion away from the atom. It is clear that the β ray continues the motion of the γ ray to a greater or less extent, and we should like to know the exact amount of this tendency. We can imagine the chance that the secondary β ray will leave the atom in any specified direction to be expressed as a function of the angle which that direction makes with the direction of the γ ray. We shall no doubt be able eventually to determine this function by experiment, and we shall have to use thin sheets of absorbing material for the purpose, for the same reason as in the case of β rays (p. 88). With thick sheets the results must be complicated and difficult to interpret, since it is then necessary to take into account the production of β rays in each stratum crossed by γ rays and their subsequent scattering and slowing down. Work of this kind has scarcely been attempted. The results given in Table XIX. can be employed only in forming an estimate of the relative numbers of β particles that go forwards and go backwards respectively from the atoms struck by the γ rays, the terms having reference to a plane perpendicular to the direction of the exciting rays. It appears from calculation that they must nearly all go forward, and it is easiest to assume this as a hypothesis and see how closely it explains the experimental results. We shall also be obliged to make certain other assumptions.

We must assume some relation between the "absorption" of γ rays and the quantity of β rays produced. The simplest supposition is that the two are always proportional to each other; that is to say if a certain percentage of a γ ray stream disappears in passing through a plate, the same quantity of β radiations makes its appearance no matter of what substance the plate is composed. As a matter of fact, it will appear later that probably the whole of the disappearing γ ray energy reappears as β ray energy, the material in which the β rays arise being simply the cause of a transformation of energy, but we do not assume so much as this at present.

Next let us suppose ourselves to be using γ rays of great penetration, which Wigger, McClelland and others have shown to be absorbed fairly strictly according to a density law: which means that screens of equal weights absorb equal amounts no matter what the material of the screens; lead only and substances of the largest atomic weights have a little more absorbing power than the rest.

Again, let us assume that the β rays are also absorbed according to a density law. In doing so we make a considerable departure from experimental results; but it is easy to allow for this error afterwards.

Finally, let us assume that the β ray when first produced continues exactly in the line of motion of the γ ray.

We proceed to compare the quantities of β radiation which should emerge from the "emergence" sides of two plates of different materials. Let these be represented by AD and $A'D'$ in Fig. 48, and let BC and $B'C'$ be corresponding strata of equal weight; in fact, let $AB/A'B' = BC/B'C' = CD/C'D' = \rho'/\rho$, where ρ and ρ' are the densities of the two plates respectively. Let the plates be crossed by equal pencils of γ rays as shown by the dotted lines in the figure. Since the same quantity of γ radiation is absorbed in BC and $B'C'$, the same quantity of β radiation takes its rise in each. And, since the strata CD and $C'D'$ are of equal weight, the same fraction of this β radiation passes out of each plate. Integrating for all effective strata, the whole emergence β radiation should be the same for each plate and so for all plates, presuming of course that the γ stream issuing from the plate is always the same.

If we now take into account the fact that the absorption coefficients of the β rays are not all equal, but diminish considerably in the case of the lighter atomic weights, then the emergence radiations should decrease as the absorption coefficients and the atomic weights increase. H. W. Schmidt (*Jahrbuch der Rad. und Elek.*, vol. v., 4, p. 486) gives the following values of the absorption coefficients of Ur β rays :—

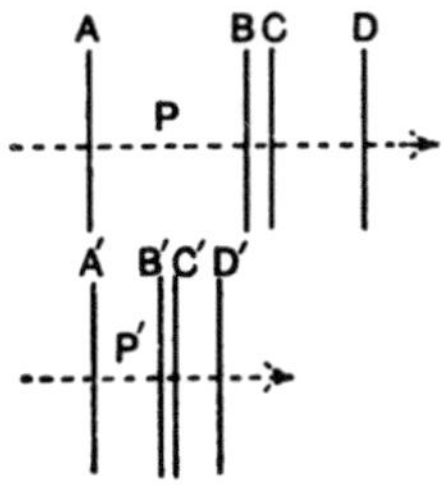

Fig. 48.

Al 5·66, Fe 7·32, Cu 7·39, Zn 7·31, Sn 7·95, Pb 9·12.

Crowther (*Phil. Mag.*, xii., p. 379, 1906) finds the values to be

C 4·4, Al 5·26, Fe 6·4, Cu 6·8, Zn 6·95, Sn 9·46, Pb 10·8.

If we take into account also the somewhat high absorption coefficient of Pb for γ rays, we see that the relative values of the figures in the last column of Table XIX. are quite in agreement with what we should expect. The lightest atoms give the largest emergence radiation, but lead has a rather high value which breaks the general rule.

The whole of this comparison is of course approximate only; the assumptions made are many and the method of calculation is a rough one.

Let us now consider the β rays appearing on the incidence side. Of those originating in BC and continuing at first the direction of the γ rays, a certain fraction, say p, is returned from the material in front, CD. The values of p for different substances have been measured by Eve, McClelland, and others. Its meaning, as already mentioned, is anything but definite, but is clear enough for present purposes. In the case of the other stratum, $B'C'$, the amount returned is p'. The same fraction of the returned portion emerges from the plate in each case, and integrating for all effective strata the incidence radiation of each substance should be proportional to its corresponding constant, p. In other words, the incidence β radiations should bear nearly the same relations to each other whether β rays or γ rays are the primary radiation. This has already been proved experimentally by Eve and McClelland: see the table on p. 119. The experimental fact is now a deduction from the hypothesis which we have assumed.

In this comparison of the incidence radiations, we ought no doubt to make the same allowances for the want of accuracy of some of the assumptions as we did in the case of the emergence rays The absorption coefficients of the β rays in C and Al are less than in the heavier atoms, and on that account the experimental values of the incidence radiations for those substances should be somewhat greater than the calculated. But it is scarcely worth while to look carefully into these sources of error and others which are also present, such as the variations in quality of returned β radiations. We do not know them with sufficient accuracy to make proper allowance for them; it is clear only that they cannot interfere with the general agreement between theory and experiment. We can conclude that the secondary β radiation, at least when originating in the lighter atoms, starts off almost entirely in the direction of the primary γ radiation to which it is due. In the case of the heavier atoms this may not hold so well.

There is one apparent discrepancy in the figures of

Table XIX. which must not be overlooked in the attempt to give a general explanation. In some cases, the emergence radiation appears to be a little less than the incidence. This is probably due to experimental error and to defects in the arrangements and the calculations, for which insufficient allowance has been made. Some of these have been mentioned already.

Some of the parallel laws have been proved by Sadler for the β rays due to X rays (*Phil. Mag.*, March, 1910). Barkla has shown that when primary X rays of sufficient

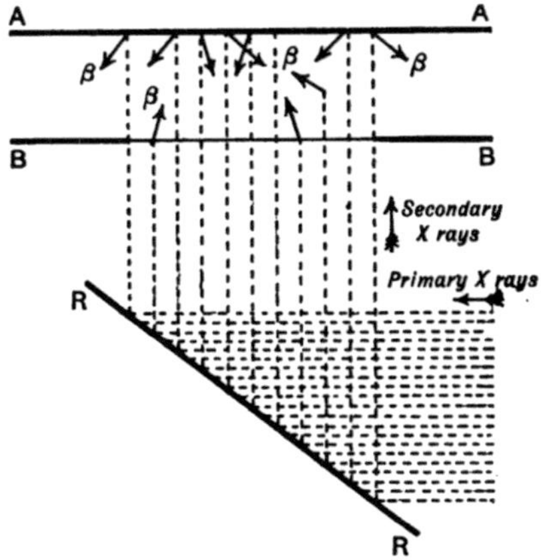

Fig. 49.

penetrating power fall upon plates of various substances, most conveniently those between chromium and tin, there is a strong secondary X radiation which is homogeneous and characteristic of the substance. The great variety in the quality of these radiations is shown by the following list of mass absorption coefficients in aluminium given by Barkla and Sadler (*Phil. Mag.*, May, 1909): Cr 136, Fe 88·5, Co 71·6, Ni 59·1, Cu 47·7, Zn 39·4, As 22·5, Se 18·9, Ag 2·5. With this range of quality, it is possible to make a searching test of the principle that the speed of the secondary β ray depends on the quality of the radiation but not upon the atomic weight of the substance.

The method will be understood from the accompanying

diagram (Fig. 49) of the essential features of the apparatus. The secondary rays characteristic of some radiator R enter the ionisation chamber AB through a thin Al sheet stretched across an opening in B. After crossing the chamber they strike the opposite wall A. The β rays from both A and B contribute to the ionisation in the chamber, the former being usually in greater quantity by far. The plate B can be moved so as to make the depth of the chamber equal to any desired value. When it is large and the β rays from the walls cannot cross the chamber, the

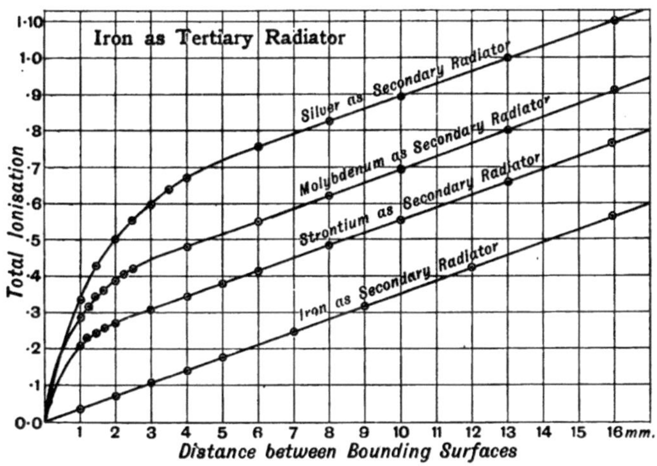

FIG. 50.

ionisation current increases uniformly with the width. But when B is pushed so closely up to A that the β rays cannot complete their paths within the chamber, the current falls off at an accelerating rate as the depth diminishes. Figs. 50 and 51 are taken from Sadler's paper (*Phil. Mag.*, March, 1910).

The abscissæ in these figures are the distances between the two plates of the ionisation chamber: the ordinates are the observed ionisation currents. The curve marked "silver as secondary radiator" refers to experiments in which the *X* rays used were the homogeneous rays emitted from silver when irradiated by the primary rays from the *X* ray bulb. Let us refer to them as Ag *X* rays. These rays were passed into the chamber *AB* (see above) through the thin Al wall in *B* and fell upon *A*, which was in this case an iron plate—the "tertiary radiator" of the figure. The curve shows that beyond a certain point the observations when plotted lie on a straight line. Sadler assumes that the increasing values of the current are now

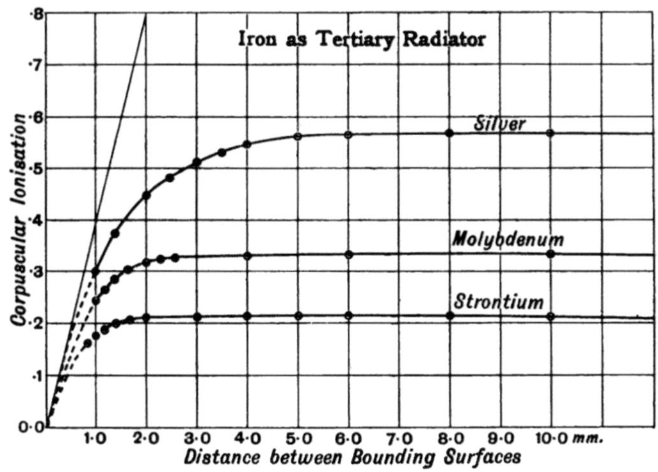

FIG. 51.

due entirely to the increasing width of air which is submitted to the direct action of the *X* rays crossing the chamber, and that a straight line drawn through the origin parallel to this straight portion of the curve will represent the direct effects of the *X* rays at all pressures. If the ordinates of the straight line through the origin be subtracted from those of the curve, the remainder can be ascribed to the action of the β rays from the walls. The curve thus obtained is given in the "silver" curve of Fig. 51. Sadler finds this curve to be very nearly of the form $y = A(1 - e^{-\lambda x})$, and he takes the value of λ so found to be a measure of the absorption coefficient of the β rays in air.

Some of Sadler's results are not in complete agreement with those of other observers. But as regards the relative penetrating powers of the β rays excited in different radiators by X rays of different qualities, his methods seem quite valid and his conclusions have been confirmed by others. Sadler puts his results into the form of a table, which is here reproduced :—

TABLE XXIII.—*Values of the Absorption Coefficient* (λ) *in Air of the Cathode Rays due to X rays of Various Qualities.*

Tertiary Radiator.	Radiators which act as the source of homogeneous secondary X rays.									
	Ni	Cu	Zn	As	Se	Sr	Mo	Rho	Ag	Sn
Iron . . .	38·9	37·0	35·8	30·2	26·4	21·5	15·5	10·9	8·84	6·41
Copper .	...	...	36·2	30·4	. .	20·8	15·2	10·8	8·81	6·67
Silver . .	...	...	35·4	30·2	...	21·2	15·4	10·3	8·78	6·63
Aluminium	...	...	...	29·6	...	20·0	15·2	.	8·90	6·54

These results show very clearly that the same principles hold in the case of the β rays due to X rays as were found by Madsen and myself in the case of the β rays due to γ rays. The penetrating power of the β ray, and its speed, depend upon the quality of the exciting ray and not upon the nature of the atom in which it is excited.

Beatty has also done independent work in this direction (*Proc. Camb. Phil. Soc.*, xv., p. 416, 1910); he used only one metal—silver—as the source of the β rays, but so far as they go his results, like Sadler's, are quite in accord with the principles we have stated.

It will be observed that the methods employed in the

investigations with X rays are different in character from those used with the γ rays. This is necessarily the case. The β rays due to γ rays are so penetrating that they only spend a fraction of their range within the ionisation chamber as it is usually constituted, and it is practically impossible to vary the pressure of the air or the depth of the chamber so as to include the whole range; we cannot therefore find the form of the whole of the absorption curve in air. We must vary the thickness of the chamber wall through which the β rays enter and so find the absorption curve in the material of the wall. In the case of the β rays due to X rays, the thinnest metal leaf absorbs completely all but the fastest, and the only way is to find their absorption in the gas of the chamber. Beatty does this by varying the pressure, Sadler the depth of the chamber. This latter method was first used by Townsend some years ago (*Proc. Camb. Phil. Soc.*, x, p. 217, 1899). Townsend's rays were heterogeneous, and his results therefore of less value than those of later date, but it is apparent from his figures that the penetrating power of the β ray is independent of the nature of the atom in which it arises.

The relations between the directions of the β ray and the exciting X ray have been investigated by Cooksey (*Nature*, April 2nd, 1908), who showed that the same want of symmetry of the β rays existed as had previously been found in the case of the γ rays. He found the emergence β radiation to be 50 to 90 per cent. greater than the incidence in such cases as he examined. He used heterogeneous X rays. He afterwards repeated the experiments with X rays of various qualities (*Nature*, Dec. 2nd, 1909), and showed that the dissymmetry increased with the penetration of the X ray. Beatty found the following values for the ratio of the emergence to the incidence β radiation (*Proc. Camb. Phil. Soc.*, xv., 6, p. 492, 1910):—

TABLE XXIV.—*Ratios of Emergence to Incidence β Radiation.*

I. Radiator.	II. Ag.	III. Cu.
Fe	1·02	...
Cu . . .	1·01	..
Se	1·10	1·08
Ag	1·29	...
Sn	1·303	1·319
Al . . .	1·435	1·42

The first column gives the metal used as the source of secondary X rays, and the radiators are arranged in the order of the penetration of the rays which they emit. Those from Fe to Sn give their characteristic radiations; Al merely scatters primary rays which are harder than any of the secondary X rays. The second column gives the values of the ratio of emergence to incidence β rays where silver is the substance in which the β rays are excited. The third column gives the corresponding values for copper.

The following values of the ratio were found by Porter and myself, using Sn X rays (*Proc. Roy. Soc.*, 85, May, 1911):—

TABLE XXV.

Metal in which β rays were excited.	Ratio.
Al	1·80
Fe	1·50
Ni	1·50
Cu	1·50
Sn	1·36

These figures give an idea of the amount of the dissymmetry. It is much smaller than in the case of the β rays excited by γ rays.

Not only do the experiments with γ rays and with X rays establish separately the principles stated, but also we find further confirmation when we take the two sets of results together; for the X rays and the γ rays may be considered as extremes in the matter of quality, and we find accordingly that the velocity of the β ray due to the γ ray is much greater than that of the β ray due to the X ray.

We may make a final statement of the conclusions to be drawn from all these experiments as follows:

1. The speed of the β ray due to the γ or X ray depends only on the quality or penetrating power of the exciting ray. The speed and penetration of the former increase with the penetration of the latter. The speed depends neither on the intensity of the γ or X ray stream, nor upon the nature of the atom in which the β ray arises.

2. The initial direction of the movement of the β ray is more or less in continuation of that of the γ or X ray, this effect being most pronounced when the exciting ray is penetrating and the atomic weight small. In the case of hard γ-rays and light atoms, the continuance is almost complete; in the case of soft X rays and heavy atoms, it is very small.

We can now proceed to consider the form which these conclusions would lead us to assign to the γ and the X ray.

CHAPTER XIII

THE CORPUSCULAR FORM OF THE *X* RAY.

THE facts which we have just established are very significant when we consider the source of the energy of the β ray. We may suppose :—

(a) That the energy comes from the atom, the X ray "pulling the trigger" and causing the release of atomic energy.

The facts do not fit this theory, for if it were true we should expect every atom to have its own particular rate of expulsion of the β ray. In the case of the radioactive substances, which expel their radiant quantities at the expense of their own energy, no two of them give off rays with the same initial velocity. On the other hand, we should not expect the velocity of the β ray to depend on the quality of an agent that merely pulled the trigger. And, further, we should not expect the direction in which the β ray first moves to depend on the direction in which the γ or X ray was travelling when it pulled the trigger. It is to be remembered also that the radioactivity of such substances as are primarily radioactive cannot be accelerated or delayed by any known physical or chemical agency, including the action of radioactivity itself. Radium does not break up any the more quickly when it is concentrated; it does not part with its own energy even when thoroughly riddled with its own radiations any quicker than when it is left alone.

In order to make this hypothesis useful it would be

necessary, among other things, to suppose that every atom contained an infinite number of systems, one to respond to every variety of X or γ ray, and perhaps an infinite number of systems of each kind, of which only such would respond as pointed more or less in the right direction. Or, at least, it would be necessary to imagine all atoms to go through stages, continuously and successively, each one of which would correspond to one of these systems. We cannot take so complicated an idea as the basis of an hypothesis unless we can find nothing simpler.

Let us therefore pass on to the alternative, and suppose

(*b*) That the energy of the β ray is derived from that of the X ray.

This may be conceived as happening in either of two ways. We may imagine an atom to collect energy from many X rays until it has enough for one β ray; or we may imagine a single X ray to furnish the necessary energy in one act. Let us take them in turn.

The first assumption is naturally associated with any hypothesis which supposes the X ray to spread its energy over a wider and wider surface as it moves away from its source. It is easy to show that in such circumstances the energy becomes so diffused that an enormous number of X rays must pass over any atom before it can gather enough for one β ray.

An X ray stream of ordinary quality loses about $\frac{1}{500}$th part of its energy in crossing one cm. of air. Consider an X ray which has spread to a distance of 20 cm. from the anti-cathode so that its energy is distributed over an area of $4\pi \times (20)^2$ sq. cm. While the radius increases to 21 cm. the X ray sweeps over $4\pi \times (20)^2 \times 3 \times 10^{19}$ molecules of air and loses $\frac{1}{500}$th of its energy. All the molecules act separately on the X ray, and we will assume that they all act similarly, each taking energy from the ray to the same extent. It follows that at a distance of 20 cm. from the origin of the ray a molecule of air can only receive

$1/(4\pi \times 20^2 \times 3 \times 10^{19} \times 500)$ or $1/7{\cdot}5 \times 10^{25}$ of the energy of the ray. The energy of the *X* rays emitted by a bulb is far less than the energy of the cathode rays inside the bulb; and the energy of the single cathode ray in the bulb is not much greater (see later) than the energy of the secondary β ray. Thus at least $7{\cdot}5 \times 10^{25}$ electrons must be passed across the bulb before a single molecule at 20 cm. distance can accumulate as much energy as is required for one β ray. Assuming that each electron carries a charge of $1{\cdot}7 \times 10^{-20}$ *E.M.U.* the whole charge amounts to $1{\cdot}25 \times 10^6$ units, which is far more than any tube takes in the course of its life. But experiment shows that the β rays make their appearance the moment the bulb is put into action.

If, therefore, we maintain this hypothesis we must suppose atoms to be in all stages of loading, some being so nearly full that they discharge their β rays immediately, while others are filled up in turn. There must be in every gas atoms which are nearer than by any assigned limit to the fullness necessary for discharge; yet since substances are not generally radioactive no atom must ever emit a β ray until the appropriate γ ray adds the last fraction of energy and precipitates the explosion. Moreover, every different quality of *X* and γ ray must be provided for. Every gas must have its set of storehouses in the proper stages of graded fullness for every kind of *X* ray over all the wide range of qualities with which we are acquainted, storehouses that will liberate a β ray of a given speed when filled up by *X* rays of the corresponding quality. The nature of the atom is not to count at all; the storehouses are to be the same in all atoms. Nevertheless, the atoms are to differ from each other in ways which will explain the remarkable variations in their absorbing powers. Finally, the greatest difficulty of all lies perhaps in this, that the direction of projection of the β ray is so greatly dependent on the direction of the *X* ray to which it is due. It is extremely hard to see how the X ray which adds the last infinitesimal amount of energy can in doing so determine the direction in which the electron shall proceed with all the energy that comes from so many X rays and has been so long in accumulating.

We see, therefore, that great difficulties lie in the way of our acceptance of the first of the two suppositions we are now considering. They arise from giving the atoms parts to play as storehouses of energy: and this in turn was rendered necessary by supposing the X rays to spread and to diffuse their energy.

Let us then turn to the second supposition: let us suppose the energy of *one* β ray to be provided by *one* X ray. All our previous difficulties disappear at once.

Because the energy of the X ray is the source of the energy of the β ray, the speed of the latter depends only on the quality of the former and is independent of the nature of the atom in which the transformation takes place; the atom is now no more than a transforming agent. The dissymmetry in the direction of projection of the β ray is explained with equal ease, for if the energy of the one ray is derived from that of the other, it can be no matter of surprise if momentum is handed over also.

Accepting this conclusion we see that the energy of the X and γ ray cannot spread as it moves away from its source. Innes's experiment has shown that the speed of the β ray does not depend upon the distance which the X ray has travelled before the transformation takes place; and therefore the X ray must retain its form, its energy, and all its characteristics unchanged as it proceeds. In a very real sense X radiation is a "corpuscular" radiation consisting of entities or quanta, each of which moves uniformly in a straight line without change until in some encounter with an atom the X ray energy disappears and β ray energy takes its place.

It is now a matter of great interest to examine the

correlated phenomenon in which the β ray dives into the material of the anti-cathode and is lost there, while the *X* ray takes its place. Just as the speed of the secondary β ray depends on the quality of the *X* ray to which it is due, so also in the converse process the quality of the *X* ray depends on the speed of the β ray which is its origin. For it is well known that the greater the potential that can be applied to the *X* ray bulb, and therefore the faster the cathode ray, the more penetrating is the *X* radiation. Moreover, in spite of apparent contradictions, the quality of the *X* ray is independent of the nature of the atom in which it arises, though if a heterogeneous bundle of cathode rays is thrown upon an anti-cathode the material of the latter may be selective in its response to the cathode particles of different velocity, and the quality of the resulting *X* radiation will then vary with the material. To complete the parallelisms it has been shown by Stark (*Phys. Zeit.*, 1909, p. 902) that the direction of the *X* ray is related to that of the cathode ray which causes it, the two directions being more or less continuous, especially when the weight of the atom is small. Kaye (*Proc. Camb.*, xv, 3, Aug., 1909) has also come to the same conclusion, but, as Stark points out, his method is not entirely free from objection.

We must also remember that the speed of the β ray which causes the *X* ray is of the same order as the speed of the β ray which that *X* ray afterwards produces. Taking all these considerations together, we adopt the hypothesis which fits them all and suppose that *one* β ray goes to the making of *one* *X* ray. The process of the *X* ray then takes on a very simple form. The electron in the *X* ray bulb is driven by the applied electromotive force against the anti-cathode, and in encounter with some atom therein loses all its energy, which is transferred to what we may call an *X* ray corpuscle. This moves away, passes, it may be, through the wall of the bulb, through the air outside and other materials, but is at any time liable to disappear in its turn as the result of some special encounter with an atom through which it happens to be passing. The energy is once more handed over to a new form of carrier, the secondary β ray. Not much energy, if any, is lost at either transformation; the exact amount is a proper subject of inquiry. We are dealing—ignoring for the time the possibility of this loss—with one and the same quantity of energy throughout, which moves in entire independence of all similar quantities contained in the energy stream. The β ray and X ray are, as it were, interchangeable corpuscular forms, the one well known in regard to its principal properties, the other new and to be investigated.

In the chapters that follow, I shall assume the accuracy of the conclusions just reached, and adopt a corresponding phraseology, speaking of the X ray as a corpuscle. It is not necessary to attach as yet any meaning to the term corpuscle, other than that which the conclusions require. The X ray is a definite entity or quantum, which moves uniformly in a straight line with unchanging form and characteristics until some action from without causes a change. I shall first of all consider some deductions which flow from this conception, and examine the extent to which they have been verified.

We know, for example, that a β ray loses energy in passing through matter; but it is now clear that an X ray cannot do so. If it is to be able to move for any distance, however great, and still keep its whole energy ready for transference to the β ray when at last the fatal encounter occurs, it cannot spend energy as it goes. We can scarcely therefore, expect it to ionise a gas through which it passes, for we have no experience of an ionisation which is brought about without an expenditure of energy.

Consequently, the ionisation which is found in a gas when traversed by X rays, must be due to the β rays to which the atoms of the gas or of the chamber-walls have brought about a transference of *X* ray energy, (*Phil. Mag.*, Sept., 1910, p. 397). In the gas itself, the β rays must commence their movements at points irregularly distributed in the track of the *X* rays, and move over paths of the order of one or two millimetres in length, ionising the gas as they go (see, for example, Fig. 3, p. 407, of the paper quoted).

Recent experiments by C. T. R. Wilson (*Proc. Roy. Soc.*, lxxxv, May, 1911) support this deduction. By a modification of his ingenious fog method, he has succeeded in rendering the paths of ionising agents visible to the naked eye. He uses a very shallow ionisation chamber with transparent walls, the air within being completely saturated with water. He allows the ionising agents to act on the air for a moment, and immediately afterwards causes the air to expand suddenly. The ions along the tracks over which the actual ionising agents have passed have very little time to diffuse before moisture settles upon them and renders them comparatively motionless. The chamber is suitably illuminated and for a moment the fog is plainly seen to be distributed in lines, the sharper lines belonging to rays which went their way just before the expansion, and the more indefinite to those of a little earlier date. Figs. 52 and 53 show the effects obtained with α rays and with *X* rays. In the former, the characteristic rectilinear motion is revealed with beautiful definition. In the second, it is to be observed how, as our considerations have led us to expect, the tracks of the cathode rays begin as it were without warning, and extend the few millimetres which we know to be their limit of penetration. Their paths are marked by sudden deflections; if they point towards us they are foreshortened and appear more as bright spots.

These photographs give us an excellent idea of the way in which the ionisation in a gas is distributed, and in the case of the X ray photograph show the way in which the secondary β rays play their part. They do not, however,

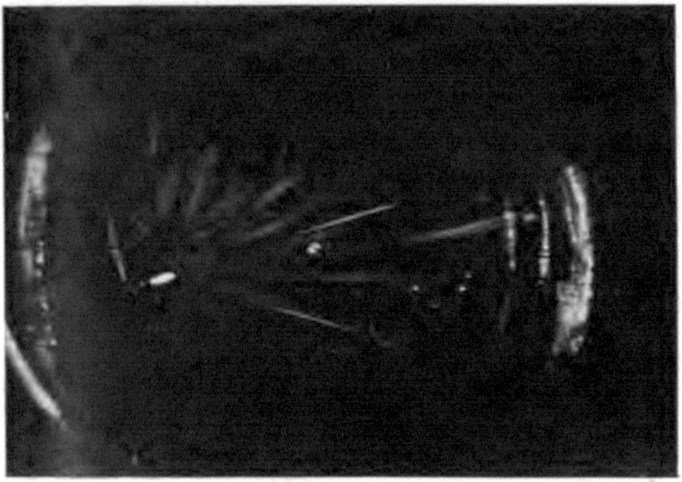

FIG. 52.—Tracks of α rays.

FIG. 53.—Tracks of cathode rays produced by X rays in air.

show that the X rays play no part at all, for a fog due to a general ionisation uniformly distributed over a plate might be invisible in comparison with the intense streaks due to the β rays.

A quantitative test has been carried out by Porter and myself in a different way (*Proc. Roy. Soc.*, lxxxv, May, 1911), the principle being as follows.

A pencil of *X* rays of definite quality is passed along the axis of a shallow cylindrical ionisation chamber. The "Sn *X* rays" (see p. 130) are most convenient to use for this purpose, because they can be obtained of good strength and because also their β rays have considerable penetration as cathode rays go (see Table XXIII.): they are able to penetrate two or three millimetres of ordinary air. The object of the experiment is to discover whether the ionisation which is observed can be all ascribed to the cathode rays which spring up in the gas of the chamber, or whether there is a remainder which must be due to a direct action of the *X* rays themselves.

Now we cannot find directly the amount of ionisation due to the cathode rays made in the gas, but we can find it indirectly; for, according to the hypothesis which we have adopted, the absorption of *X* rays implies a conversion of *X* ray energy into β ray energy, and we assume, as we have done previously, that the β ray energy which appears is at least proportional, even if we cannot yet suppose it to be equal, to the *X* ray energy which disappears, no matter what the substance which causes the transformation. As we shall see presently, we can find the ionisation in the gas due to the absorption of the *X* rays by a given weight of metal—silver foil is here used—and thence we can calculate the ionisation in the gas due to the absorption of the *X* rays by the gas; in both cases, through the direct action of β rays.

The apparatus is shown in Fig. 54. A primary pencil of rays from an *X* ray bulb passes down a tube lined with tin and falls upon a plate of tin in a box made of tin. The primary rays fall on nothing but tin, the object being to make the pencil of rays which issues from the box as free as possible from all but Sn X rays. The secondary rays pass into an ionisation chamber through an opening AA closed with paper. They excite emergence β rays in sheets of metal laid on the paper, or incidence β rays in plates laid on top of the chamber when such sheets or plates are placed into position. The ionisation current is measured by balancing it against an opposing current due to the ionisation in a second chamber crossed by a pencil of X rays from the same radiator. The balancing chamber has a window with an adjustable shutter worked by a micrometer screw, and the values of the readings are determined by a special calibration. A Wilson tilted electroscope is used as a detector. The balance can be found quite easily with an accuracy of less than one per cent.

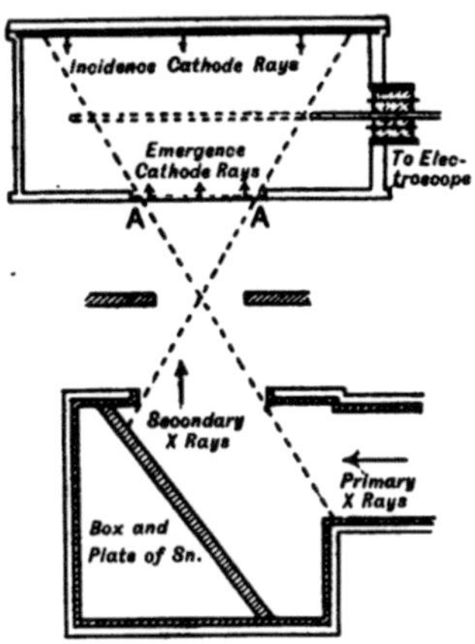

Fig. 54.

The whole of the chamber, which is made of brass, is lined with aluminium and finally with paper, so that there is no appreciable amount of secondary X radiation, nor is there any special cathode radiation to be taken into account. This point is fully discussed in a later chapter. The current is first measured when the X rays enter the chamber through paper only, and then silver foils are laid one by one on the card, and the increase is observed as the number grows. The results are shown in Fig. 55. If we can measure the slope of this curve at the origin, we shall know the value of the cathode radiation for very thin sheets, and this is what we really require, for it represents the effect of the *whole* of the cathode radiation which is produced. When the sheets are vanishingly thin we can suppose all the cathode rays made in the sheet to get out into the gas and exercise their full effect.

Unfortunately, silver foil is not to be had thin enough to give any points nearer the origin than are shown in the figure, and we are left to draw the curve through the experimental points as best we can. We may of course assume that the curve is of the form $y = A(1 - e^{-\lambda x})$ as in the parallel case in Chap. XII, p. 120. The curve is not quite exponential, however, and therefore it was assumed by Porter and myself that it would be best to take the simple form $y = ax - bx^2$ and to find the constants a and b, the former

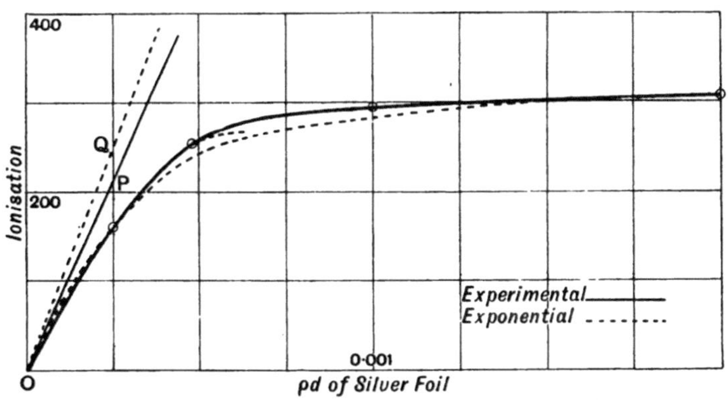

FIG. 55.—Cathode ray ionisation in terms of the weight (ρd) in gr. per cm² of the silver foil through which the *X* rays have passed.

of which is actually the quantity wanted. The tangent to this curve is *OP* in the figure; the curve passes through the first two points, of course, and then breaks away as shown. The complete dotted curve is an exponential one drawn so as to pass through the point on the experimental curve for which the current is half its final value. The tangent to this curve at the origin is *OQ*, and this represents the result of fitting an exponential curve to the experimental determinations as it is often done.

From the inclination of the curve at the origin as shown by *OP*, it appears that when the Sn *X* rays pass into the chamber through a silver leaf weighing 0·00025 gr. per sq. cm. the cathode rays produced in this sheet and passing onwards towards the chamber would, if no part of their energy were abstracted in crossing the leaf, cause an ionisation measured by 208 on the arbitrary scale of the micrometer screw (duly calibrated). It should perhaps be pointed out, for the sake of clearness, that the strength of the primary rays does not affect such numbers as these, for the experiment balances the effect due to one stream of secondary X rays against another stream of similar rays derived from the same source.

In order to find the full measure of the cathode radiation produced by the absorption in the silver, we must also take account of that which would be found on the incidence side of the foil. We ought really to do the same thing in regard to incidence cathode rays as we have just done in regard to emergence. But it is simpler and more accurate to compare the incidence and emergence radiations by laying one or more silver foils on a card and placing this in the middle of the ionisation chamber, so that the X rays pass through it. The two radiations are then found by subtracting the current for a bare card from the current when there are silver foils on one side or the other. We found the ratio between emergence and incidence radiations to be 1·30 for one, two, or three foils: and we conclude that this ratio must hold for extremely thin films. It follows that the distribution of the cathode rays about the atom is not dependent—for these small thicknesses of silver—upon any scattering of the cathode rays by the material. The cathode rays do not continue the motion of the exciting X rays to anything like the same extent as do the β rays caused by γ rays.

The measurements so far recorded were made in air, because it was convenient to adjust the silver foils in a chamber which could be easily opened and closed. The general conclusion can, however, be made independent of the nature of the gas and of the form of the chamber employed by putting it in the following way:—

When the Sn X rays cause an emergence β radiation from silver measured by 309 (see Fig. 55), the same rays cause an actual production per milligramme of silver (*i.e.* four times the weight of the silver foil used in the experiment) measured by $4 \times 208 = 832$ for emergence β radiations alone or $832 \times 2{\cdot}3/1{\cdot}3 = 1470$ for all β rays.

The remaining part of the experiment was carried out with an ionisation chamber full of oxygen, for the method requires a knowledge of the relative absorption coefficients

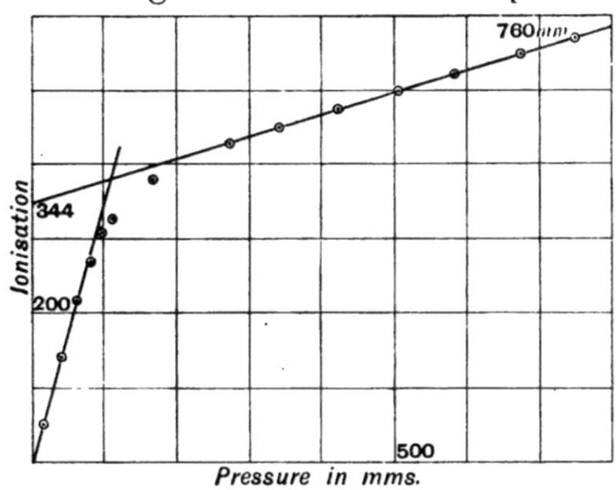

FIG. 56.—Ionisation current in terms of the pressure of the oxygen in the chamber.

of the Sn X rays by the silver and by the gas. It is difficult to find the absorption coefficient of X rays in air, or any substance which must be handled in the form of a gas, but the absorption coefficient in oxygen can be found by the use of solid absorbing screens containing known quantities of oxygen.

The Sn X rays were passed into an ionisation chamber which could be filled with oxygen at any pressure. They entered through a screen of ten silver foils, which was amply sufficient to give the full emergence β radiation. The depth of the chamber was 3·45 cm. The results are shown in Fig. 56, where the ionisation observed is plotted against the pressure of the oxygen. The upper portion of the curve is rectilinear, and is produced backwards to cut the vertical axis, which it does at a point having an ordinate 344. This figure must be taken as a measure of the ionisation in the gas due to the emergence cathode rays from the silver. At a pressure of 760 mm. the ionisation is 571, so that the effect of the X rays upon the oxygen at that pressure is $571 - 344 = 227$.

If the emergence β radiation had been measured by 309, the Sn X rays would have been of such strength as to have caused an ionisation of $227 \times 309/344 = 204$ in the layer of oxygen. The latter was 3·45 cm. wide, and the density of oxygen at the pressure and temperature of the experiment was 0·00137. The same rays would therefore have caused an ionisation of $204 \times 1/3{\cdot}45 \times 1{\cdot}37 = 43{\cdot}2$ in a milligramme of oxygen. As shown above, such a pencil of Sn X rays would have produced from one milligramme of silver enough β radiation to cause an ionisation in oxygen measured by 1470.

The absorption coefficient of Sn X rays in silver was found to be 15·4, the circumstances of the experiment being exactly the same. The very soft constituent of the Sn X rays was intercepted by a wooden plate which closed the ionisation chamber, and was used throughout these experiments because it stood air pressure and did not cut out the harder part of the Sn X radiation. The absorption coefficient of carbon was found to be 0·333, and that of oxalic acid 0·388. Beatty has shown that the absorption coefficient of hydrogen is negligible (*Proc. Camb. Phil. Soc.*, xv, 5, p. 422, 1910). Since oxalic acid with its water of crystallisation contains one carbon atom to three of oxygen and three of hydrogen, we can calculate the absorption coefficient of oxygen from the equation :

$$12 \times 0{\cdot}333 + 48x + 3 \times 0 = 63 \times 0{\cdot}0388, \text{ so that } x = 0{\cdot}425.$$

Finally, therefore, the pencil of Sn X rays which we are taking as the standard of comparison could produce in a milligramme of oxygen enough β rays to cause an ionisation of $1470 \times 0{\cdot}425/15{\cdot}4$ in the same gas, or 40·6. The ionisation actually found was 43·2. The difference between these quantities is within the errors of calculation and experiment, and therefore the result is in agreement with the inference under test, viz. that the *X* ray does not ionise directly but only indirectly through the β ray to which its energy is transferred.

It has recently been pointed out by Dr. Barkla (*Phil. Mag.*, Feb., 1912) that the absorption coefficient of oxygen found in this way is not a true measure of the production of β ray energy, since a substance of small atomic weight scatters primary *X* rays, and some of these must miss the ionisation chamber wherever the absorbing screen is placed. This criticism is quite just. But silver also distributes primary radiation without converting it into cathode rays, and experiment shows that the two effects nearly balance. If we vary the positions of the absorbing screens, placing them, now nearer, now further from the ionisation chamber, the values of the absorption coefficient vary separately, but the ratio changes very little. When a plate of silver or a block of carbon is placed on the top of the chamber, the radiation turned back into the chamber is nearly the same in each case, both in quality and quantity. I think, therefore, that our ratio of the absorption coefficients of Sn *X* rays in silver and oxygen, *viz.* 36·3, is sufficiently accurate.

CHAPTER XIV

THE ENERGY OF THE X RAY

WE are assuming that one β ray provides the energy for one X ray when this particular type of transformation occurs: and that the one X ray in its turn provides the energy for one β ray. Since the energy of the latter β ray is of the same order as that of the former, we conclude (Bragg and Madsen, *Trans. Roy. Soc. of South Australia*, xxxii, 1908, *Phil. Mag.*, Dec., 1908, that the energies of the β ray, the X ray and the secondary β ray are all nearly the same, and that little, if any, energy is lost in either of the transformations.

The definition of the quality of X radiation is generally made in terms of the absorption coefficient of the radiation in some standard substance, and aluminium is chosen for the purpose. It has been very convenient to use this method of definition, but it is open to the objection that measurement ought not to depend in any way on the properties of an arbitrarily chosen substance. It is true that aluminium does not show any anomalies within the range of the X ray qualities usually considered, but other substances do; and there is little doubt that aluminium would if the range included were wider. It is therefore important to observe in the first place that the energy of the X corpuscle would be a more satisfactory definition of quality if it should turn out to be sufficient and ascertainable, and in the second place that these conditions appear to be satisfied.

The loss of energy at transformation is small, if it exists; it must obviously be independent of the material in which it occurs; and therefore little harm can be done in neglecting it in many investigations which we may make. We can say therefore that the energy of the X ray is ascertainable. Again, there is no evidence that two X radiations can differ in quality and yet give rise to a β ray of the same speed. Any change in the penetrating power of the X ray brings about a change in the initial speed of the β ray. We only know X rays by their energies, and these we assume to be the energies of their corresponding β rays. If any loss of energy at transformation is afterwards discovered it will be easy to make the proper allowance, and the simplicity we gain more than counterbalances the risks we are taking. The energy of the X ray is therefore a sufficient definition.

From this point of view let us proceed to consider investigations which have dealt especially with the qualities of X rays from various sources. As we have already said, we owe most to Barkla and Sadler in this field of inquiry. The accompanying table is taken from their paper in the *Philosophical Magazine* for May, 1909, p. 749. The "radiator" is the plate on which the primary X rays are allowed to fall and is then found to be the source of homogeneous X rays of definite quality, the absorption coefficients in various substances being shown in the table :—

TABLE XXVI.—*Mass absorption coefficients* $\left(\frac{\lambda}{\rho}\right)$.

Radiator	Absorber.										
	C	Mg.	Al.	Fe.	Ni.	Cu.	Zn.	Ag.	Sn.	Pt.	Au.
Cr . .	15·3	126·5	136	103·8	129	143	170·5	580·5	713·7	[516·8]	[507 +]?
Fe . .	10·1	80	88·5	66·1	83·8	95·1	112·5	381	472	340	367
Co . .	7·96	63·5	71·6	67·2	67·2	75·3	91·5	314	392	281	306
Ni . .	6·58	51·8	59·1	314	56·3	61·8	74·4	262	328	236	253
Cu . .	5·22	41·4	47·7	268	62·7	53·0	60·9	214	272	194	210
Zn . .	4·26	34·7	39·4	221	265	55·5	50·1	175	225	162·5	178·2
As . .	2·49	19·3	22·5	134	166	176	203·5	105·3	131·5	105·7	106·1
Se . .	2·04	15·7	18·9	116 3	141·3	149·8	174·6	87·5	112	93·0	100·0
Ag . .	·41	2·2	2·5	17·4	22·7	24·3	27·1	13·3	16·5	56·5	61·4

Sadler has shown that each quality of X ray produces a cathode ray of characteristic initial speed and penetration (see p. 131); Beatty has proved the same result. The speed increases with the atomic weight of the radiator, and we conclude that the energy of an X corpuscle increases with the atomic weight of the atom of which it is characteristic. Now Barkla has shown (*loc. cit.*) that the homogeneous radiation of any atom can excite the homogeneous radiation of an atom of less weight, but the converse does not occur. Zn X rays can, falling on iron, excite Fe X rays; but Fe X rays cannot excite Zn X rays. In the language of the corpuscular theory, this means that one X corpuscle may give rise to another X corpuscle of less energy, but never of more; which is after all what we should expect.

If, indeed, we consider a line of transformations we must always find the corpuscle, whatever form it may have at the moment, in possession of at least as much energy as in any subsequent form. Thus the β ray in the *X* ray bulb must possess a minimum velocity if it is to excite, it may be through intermediate steps, an *X* ray of given quality. This has lately been verified experimentally by Whiddington (*Proc. Roy. Soc.*, xlv, p. 323, April, 1911). A diagram of his apparatus is given in Fig. 57. The cathode rays from *C* pass through the opening *L* and are opened out into a spectrum by a magnetic field. The slit *O* permits the

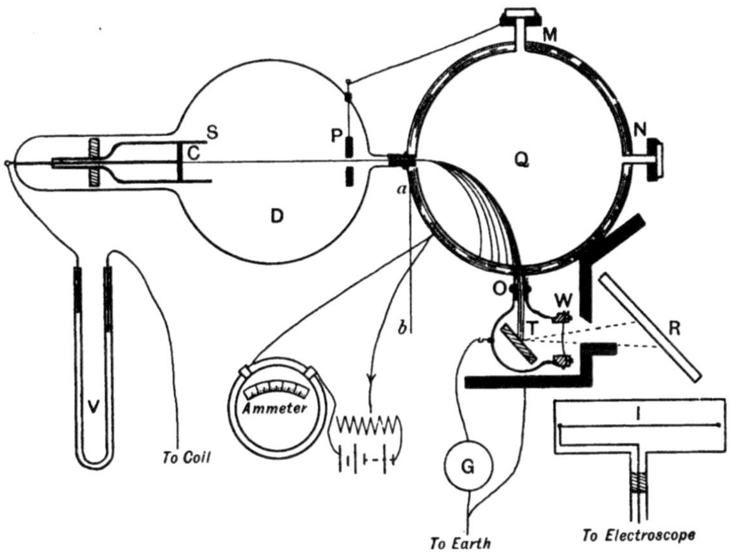

FIG. 57.—Whiddington's experiment.

passage of the rays chosen for investigation. Whiddington found that, fortunately, there was a nearly homogeneous group among the fastest rays which contained a large fraction of the whole energy of the cathode ray stream. The selected rays fall on a silver plate *T* and there excite *X* rays which pass out of the thin aluminium window *W* and fall upon a radiator *R*. The secondary *X* rays from *R* pass into an ionisation chamber *I* and the current therein is measured. The general velocity of the whole cathode ray stream is varied by a mechanical device depending on an effect discovered by Whiddington. It consists of a sliding tube S which determines the state of the bulb by the distance to which it projects beyond the cathode C (*Camb. Phil. Soc. Proc.*, 1911, xvi, p. 150).

A given material being chosen for the radiator R, the speed of the β rays falling on T is gradually increased and it is found that there is little effect in the chamber unless the speed surpasses a certain critical value; as it increases beyond this value the current grows very rapidly. The critical value (v_c) for various radiators is shown in the following table; it will be observed that it is nearly proportional to the atomic weight:—

TABLE XXVII.

Radiator.	Atomic weight.	$v_c \times 10^{-9}$.	V_c.
Al	27	2·06	1,200
Cr	52·5	5·09	7,320
Fe	56	5·83	9,600
Ni	58·6	6·17	10,750
Cu	63·2	6·26	11,080
Zn	65·1	6·32	11,280
Se	78·9	7·38	15,400

The last column gives the potential in volts which must be applied to the bulb in order to obtain cathode rays of the critical speed. The figures in this column are the relative values therefore of the energies of the different X corpuscles, assuming there is no loss of energy when the first transformation occurs; if they are multiplied by the charge on the electron in electromagnetic units and by 10^8 they express the energies directly in ergs. For instance, the energy of the Zn X corpuscle is $11280 \times 1{\cdot}55 \times 10^{-20} \times 10^8 = 1{\cdot}75 \times 10^{-8}$ ergs, accepting Rutherford's value for e, or about 7 per cent. higher if we take Millikan's recently found value.

Whiddington was unable to apply sufficient voltage to

produce Sn *X* rays in the same way without rendering the bulb unsteady and making measurement impracticable It is worth observing, however, that a not unreasonable extrapolation gives about 35,000 volts as the critical potential for Sn rays. The mass absorption coefficient in most substances of the β rays due to Sn *X* rays is known to be about 3200. The curve in Fig. 55 gives this value; and Beatty found for the linear coefficient the value 3·97 (*Camb. Phil. Soc.*, xv, 5, p. 418) which is equivalent to a mass coefficient 3230, assuming the density of the air to be 0·00123. Now Lenard (*Ann. d. Phys.* xii, p. 732, 1903) states that corpuscles (electrons) projected with the speed due to about 30,000 volts have a linear absorption coefficient of 0·005 in air at 1 mm. pressure, and therefore of 3·8 at 760 mm. This example shows a close agreement between the speeds of the electrons at the two ends of the *X* ray. The data on which it is based are rather uncertain, but more accurate knowledge can scarcely injure the main conclusion.

Barkla's table of absorption coefficients shows most remarkable numerical relations, of which no doubt some simple and illuminating explanation exists if only it could be found. Consider, for example, the absorption of various rays by nickel as displayed in the column headed Ni. The *X* rays from chromium pass through with an absorption coefficient 129, which means that they are half absorbed in crossing a nickel sheet weighing 0·7/129 or 0·0054 gr. per sq. cm. As we take the *X* rays characteristic of heavier and heavier atoms up to those given by copper, the nickel is more and more easily penetrated. To put it numerically, the Cu *X* ray is 50 per cent. more energetic than the Cr *X* ray, and its chance of passing untouched or unaltered through a nickel atom is twice as great. But then comes a remarkable change. The Zn *X* ray is less than 2 per cent. more energetic than the Cu *X* ray, and yet there is a violent alteration in the absorption coefficient and that in the unexpected direction;

the coefficient is four times as great for the Zn rays as for the Cu. The table XXVI shows many other parallel cases.

Now it appears that there are other discontinuities at the same stage of the gradual increase of *X* ray energy. If we pass pencils of Zn rays and of Cu rays across an ionisation chamber, allowing them to strike a nickel plate on the farther side, and if we adjust their relative strength until they cause equal ionisation currents in the air of the chamber, then the Zn *X* rays cause far more cathode radiation to emerge from the nickel plate than do the Cu *X* rays. The sudden increase in the absorption coefficient is accompanied by a sudden increase in the production of β rays.

Still more striking is the fact that the Cu *X* rays, though slightly more energetic, individually, than the Ni *X* rays, excite scarcely any such secondary rays in the nickel, whereas the Zn *X* rays produce them in large quantity. Sadler calculated that one-third of the Zn *X* ray energy appeared again as energy of Ni *X* rays (*Phil. Mag.*, July, 1909); a rather smaller value was found by Porter and myself (*Proc. Roy. Soc.*, lxxxv, 1911, p. 360). Thus the sudden increase in the absorption coefficient is accompanied by the sudden appearance of secondary *X* rays from the nickel.

There can be no doubt that all these phenomena are intimately connected, and that there is a complete change or discontinuity in the process of absorption as soon as the energy of the *X* corpuscle exceeds the energy of the corpuscle of the atom through which it is passing—or at least close to that point. The same statement must be true of the β corpuscle, because it is only when the β corpuscle possesses energy in excess of the same value that it is able to excite the characteristic *X* radiation of the atom.

Our knowledge of the absorption coefficient of the β particle *at* different speeds is unfortunately small; for the most part measurements are made of the penetration of β particles of various initial speeds, which is not at all the same thing. The absorption coefficient of an X ray of given energy is readily found, because a stream of such rays is absorbed exponentially in passing through a plate; no matter how much weakened the stream has become, it is the number of particles that has diminished and not the individual energy of those that remain; but the individual β ray loses energy as it goes through matter. Moreover, its power of penetrating sheets of given materials, solid or gaseous, is no doubt affected by its scattering coefficient in that material. We are therefore some way as yet from knowing as much about the β ray in connection with the problem we are considering as we know about the X ray. It is necessary to find the "length of the track" (see p. 95) of β rays in nickel as a function of the velocity. If this were done and the proper curve plotted, it would no doubt have some break in it at the point or near the point where the energy is that which is characteristic of the Ni corpuscle.

If, for example, the ordinates in the curve ABC (Fig. 58) represent energy, and the abscissæ ranges, the curve will have a discontinuity at B; the dotted line representing the energy characteristic of Ni. Above this point additions to the energy produce less increases of the range than just below it; for so long as the energy of the β ray exceeds the critical value there is the chance that the β ray will disappear and be replaced by an X ray, and this is equivalent to an additional cause of diminution of β ray energy. We might even imagine a break back at B, an increase in the energy producing a decrease in the range. But we know

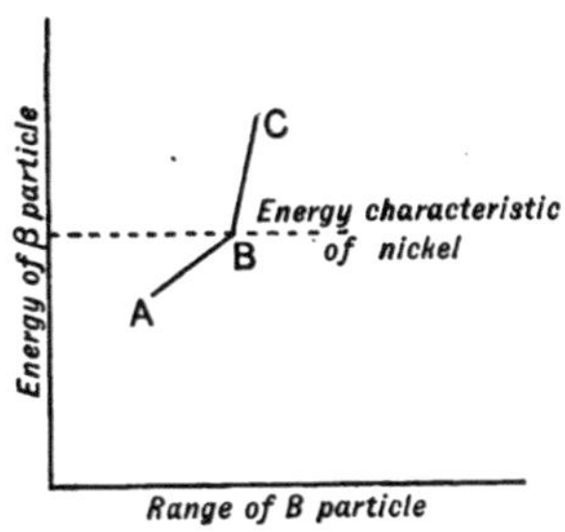

FIG. 58.

practically nothing of the relative importance of this effect; that is to say, nothing of the discontinuity of the two branches of the curve, *AB* and *BC*. I am at present engaged in an investigation of this point.

The Zn *X* ray with an energy due to 11,280 volts can give rise when it meets with a Ni atom to a Ni *X* ray with an energy due to 10,750 volts (see Whiddington's table, p. 154). If we had the information as to the absorption of the β ray which we have just been discussing, we should doubtless be able to discover where the energy represented by the difference of 530 volts went. It might be taken up by the atom in which the transformation takes place; on the other hand, it is quite possible that the Zn *X* ray is first transformed into a β ray and that the loss occurs, as we know it can do, while the energy is in this form. There is some suggestion, reviewing such facts as we have, that serious losses do not occur in single atoms. There is no great loss, if any, when a cathode ray is replaced by an *X* ray, nor in the reverse process; there is none apparently when a β ray undergoes a single deflection, though this is not quite certain, There is certainly no loss when an *X* ray is scattered, and this is probably true of γ rays also. There is some inducement therefore to adopt the second view and to suppose an intermediate β ray between the Zn and the Ni *X* ray.

Perhaps we may find some such explanation as the following. So long as the *X* corpuscle has more energy than the Ni corpuscle, it is peculiarly liable to transformation to the β ray form; and so long as the β corpuscle has the same excess, it is liable to transformation to the *X* ray form and also to loss of energy as it moves along its path, possibly *very* liable to such loss as described above. While, therefore, the excess exists there will be considerable interchange between the two forms the energy falling always while it is carried in the β ray form until it drops to the critical value, that of the Ni *X* ray. The interchanges then cease and the forms become more stereotyped. In this way a stream of Zn *X* rays will be gradually transformed into a mixed stream of β rays and of other *X* rays which will possess a certain minimum energy. How closely the secondary rays will group themselves about this minimum will depend on those properties of the β ray of which we know so little as yet.

Barkla has also arranged the absorption coefficients of Table XXVI in a second and most suggestive form, which is reproduced in Table XXVIII. All the coefficients are shown no longer in absolute measure, but relative to the absorption coefficients by aluminium.

TABLE XXVIII.

Radiator.	Absorbers.									
	C/Al	Mg/Al	Fe/Al	Ni/Al	Cu/Al	Zn/Al	Ag/Al	Sn/Al	Pt/Al	Au/Al
Cr . .	0·112	0·930	0 763	0·949	1·051	1 254	4·27	5·25	[3·8]	3·73 + ?
Fe . .	0 114	0·904	0·747	0·947	1·074	1·271	4·31	5·33	3·84	4·14
Co . .	0·111	0·887	0·938	0·939	1·052	1·278	4·38	5·47	3·92	4·27
Ni .	0·111	0·876	5 31	0·953	1·046	1·259	4·44	5·55	3·99	4·28
Cu . .	0·109	0·866	5·62	1·314	1·111	1·277	4·48	5·70	4·06	4·40
Zn .	0 108	0·881	5·61	6·7	1·408	1·272	4·44	5·71	4·12	4·52
As .	0·111	0·857	5 95	7·29	7·82	9·045	4 68	5·84	4·69	4·71
Se . .	0·108	0·831	6·15	7·42	7 93	9·24	4·63	5·93	4·92	5·29
Ag .	0·164	0·88	6·96	8·80	9·72	10·84	5·32	6·60	22·6	24·6

The uniformity of the figures in some of the vertical columns is most remarkable. The first column, for example, shows that aluminium absorbs every quality of X radiation which is dealt with in the table nine times as much as carbon. This seems to argue for some statistical explanation. We might suppose both atoms to contain similar centres of influence upon passing X rays, and that weight for weight aluminium contains nine times as many such centres as carbon. And again, if we follow the figures down several of the columns we find that the sudden increase at the critical point is always in the same proportion, about 1 to 8, as if above this point each centre became eight times as effective as it had been before.

CHAPTER XV

CALCULATION OF THE IONISATION CURRENT UNDER GIVEN CONDITIONS

We are continually measuring in these investigations the "ionisation current," that is to say, the ionisation to be expected when a given pencil of X or γ rays crosses an ionisation chamber of given form, materials, and contents. Usually the problem is a very difficult one; but the corpuscular theory simplifies the terms in which it is expressed. The ionisation is effected solely by the β rays which spring up in the path of the X rays, and the absorption coefficients of the gas and of the walls for the X rays in question provide the basis of a calculation of the quantity of β radiation which is produced. In some cases it is necessary to take into account the chances of reconversion of β rays into X rays, or the production of the homogeneous X radiation which is so important at times; in the previous chapter I have tried to show that the former of these apparent alternatives may include the second. The problem thus resolves itself into two parts: first the determination of the density and quality of the β radiation in different parts of the gas in the chamber, and secondly the determination of the ionisation which such β radiation produces in the gas.

In practice it is not possible to complete such a calculation in the general case, partly because our knowledge of the behaviour of β rays is incomplete, partly because the circumstances of the production of secondary X radiation have not been fully investigated, and partly because the calculation can become too complicated.

There are some simple cases which can be worked out with enough success to make them useful. We have already considered an example of this kind in the chapter dealing with β rays; viz., that of a chamber made wholly of one substance crossed by γ rays moving uniformly in all directions and everywhere of uniform intensity, filled with a gas unable to absorb the β radiation appreciably. We saw that in this case the ionisation varied with the length of the track of the β ray in the material of the chamber walls, and might serve as a relative measure of the length in that material.

There is another case in which a partial solution can be obtained and is very useful. When an ionisation chamber of given form is acted on by γ rays and filled with a variety of gases, the ionisation is almost wholly due to the β rays from the walls unless the gas is very dense. Consequently the different gases when placed within the chamber are subjected to the action of the same β rays. The ionisation currents should, therefore, be in the same proportion to each other as if the β rays had been used directly. Kleeman made a number of measurements of this kind with the purpose of finding the ionisations of the different gases by the γ rays; the corpuscular theory leads us to look upon the action of the γ rays as indirect only. The following table of results is taken from Kleeman's paper (*Proc. Roy. Soc.*, lxxix, 1907, p. 220); the ionisations of various gases by β rays are given in the second and by γ rays in the third column of the table. The latter were obtained by the use of the γ rays from radium, the former were due to the β rays of uranium; but as the relative ionisations due to β rays of different speeds differ so little from each other the comparison is legitimate:—

TABLE XXIX.—*Ionisations of various gases by β- and γ-rays.*

I. Gas	II. β rays	III. γ rays	I. Gas	II. β rays	III. γ rays
Air	1	1	$CHCl_3$	4·94	4·93
O_2	1·17	1·16	CCl_4	6·28	6·33
CO_2	1·60	1·58	CS_2	3·62	3·66
C_2H_4O	2·12	2·17	CH_3Br	3·73	3·81
C_5H_{12}	4·55	4·53	C_2H_5Br	4·41	4·63
CH_4O	1·69	1·75	CH_3I	5·11	5·37
$C_4H_{10}O$	4·39	4·29	C_2H_5I	5·90	6·47
C_6H_6	3·95	3·94	NH_3	0·888	0·898
C_2N_2	1·86	1·71	SO_2	2·25	2·27
N_2O	1·55	1·55	H_2	0·165	0·160
C_2H_5Cl	3·24	3·19			

The close equality of the two columns of figures is evident. The only obvious differences are in the cases of those gases which contain very heavy atoms, in which it may well be supposed that such atoms are producing a small but appreciable quantity of β radiation from the softer constituents of the γ rays.

At a later date, Kleeman extended his experiments by using the secondary γ rays excited in lead, zinc, and carbon. These are softer than the primary, as we have already seen. In this case, there was a pronounced excess of the ionisation in the gases containing the heavy atoms. A few instances are given in the following table; they are selected from the larger table in Kleeman's paper (p. 368, *loc. cit.*):—

TABLE XXX.—*Ionisation of various gases by hard (primary) γ rays and by soft γ rays (secondary from zinc).*

Gas	Primary γ rays.	Secondary rays from zinc.
Air	1	1
CO_2	1·58	1·53
C_5H_{12}	4·53	4·36
SO_2	2·27	2·17
CCl_4	6·33	6·35
CH_3Br	3·81	6·15
C_2H_5Br	4·63	6·05
C_2H_5I	6·47	12·46

Here, again, the same explanation may be given of the larger figures at the foot of the third column. It is true that Kleeman took the precaution of comparing the ionisation at different pressures in ethyl bromide (C_2H_5Br), and found it to be nearly proportional to the pressure, whereas at the lowest pressures the secondary effects of the heavy atoms in the gas ought not to manifest themselves. The curve showing the relation between pressure and ionisation cannot, however, be a straight line throughout its whole length, as will be seen later; and it is possible that there is more curvature close to the origin than Kleeman's results seemed to him to show. The ethyl iodide would have been more likely to exhibit the departure from proportionality within the range over which Kleeman worked. It is to be remembered that the conversion of the energy of soft γ-rays into that of the corresponding β rays is abnormally large for heavy atoms.

If this explanation of the divergences in the case of the heavy atoms is accepted, then over a wide range of substances the β- and γ-ray ionisations are relatively the same, as we have been led to expect.

This interpretation of results is not quite in accord with the opinions of most investigators. Kleeman was of opinion that about half of the ionisation in an air-filled chamber was due to the direct action of the γ rays upon the air. He based this conclusion on his inability to prevent ionisation occurring in an experiment which he had arranged as shown in Fig. 59.

A is an ionisation chamber 10·5 cm. long, 10·4 cm. broad, and 7 cm. deep, of which the upper and lower sides consisted of thin, tightly stretched tissue paper, equivalent in mass to a layer of air 1 cm. thick. The chamber was placed on the poles B_1 and B_2 of an electromagnet which were resting on a lead block *C*, 5 cm. thick. This lead block had an aperture *a*, 3 cm. by 3·2 cm., which was placed in a symmetrical position with respect to the poles of the electromagnet and the ionisation chamber. D is the tube containing the radium placed at a distance of 10 cm. from the lead block. The plate b was of aluminium 4 mm. thick. It was found that the ionisation in the chamber decreased to about 55 per cent. of its original value when a magnetic field of sufficient strength to prevent the β rays from the plate b entering the chamber was applied. From this and similar experiments, Kleeman drew the conclusion already stated.

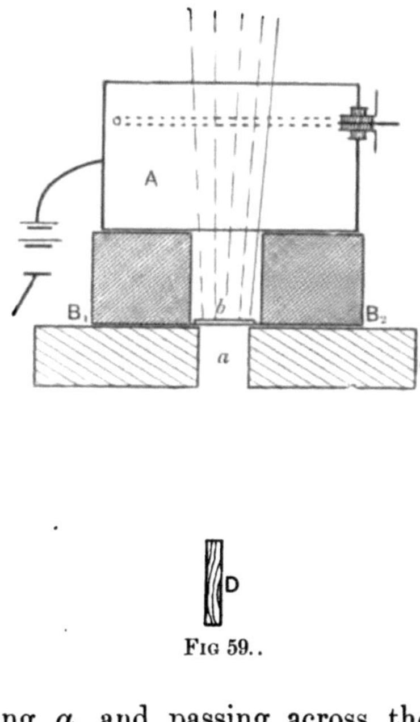

FIG 59.

The use of thin paper to form the upper and lower walls of the chamber was intended to prevent the formation of β rays in any substance except the aluminium sheet; in any substance, that is to say, which could send β rays into the chamber. It was found by Madsen and myself that such a precaution was not very efficient. The secondary radiations excited in various places by the stream of γ rays entering the chamber by the opening a, and passing across the chamber to fall upon neighbouring objects, have an effect which seriously interferes with the experiment, and causes an ionisation unresponsive to the magnetic field. Moreover, the influence of the magnetic field is not perfectly clear. It may increase a β ray effect in some ways while it lessens it in others; for that portion of a β ray path which is completed within the chamber may be lengthened by forcing it into a circular form. Also β particles are scattered by impact on atoms of the gas, and of the surfaces upon which the magnetic field deflects them, and by successive impacts may travel in spite of the field; for the field does no more than convert the rectilinear portions of the path into circular portions, and has no influence on the direction which the particles take after impact. I think the effects of magnetic fields deserve closer examination.

In another simple case the solution is obvious. If the gas in a vessel is composed of the same atoms as the walls of the chamber, or even of atoms of nearly the same weight, the ionisation must be proportional to the pressure. This is really a special case of a problem already considered, that of the ionisation chamber made wholly of one material. It has been shown in Chap. IX that the density of the β rays inside a material of uniform composition traversed by X rays uniformly distributed throughout it depends only on the characteristics of the radiation and the nature of the material; it is independent of the density of the material and has even the same value where there is a cavity. Thus if the ionisation chamber is filled with a gas of equally heavy atoms, the β ray density is independent of the pressure of the gas, and the ionisation is proportional to the pressure. Excellent examples of this effect have been given by Beatty (*Proc. Roy. Soc.*, lxxxv, 1911, p. 230).

In Fig. 60, which is taken from Beatty's paper, the lines A and C show the variation of ionisation with pressure in the case of a chamber lined with selenium and filled with SeH_2. The hydrogen atoms in the gas are of no consequence, and the gas is practically of the same material as the walls. So also for the walls B and D representing the results of similar experiments in which the chamber was lined with arsenic and filled with AsH_3. It is clear that A, B, C, and D are nearly straight lines. The ratio of the slopes of A and B is 1·50 nearly, and this shows that the absorption of the Sn rays in selenium is greater than the absorption in arsenic in the ratio 1·50 to 1. Comparing the slopes of *C* and *D*, we find that the same quantity when Sn rays are replaced by Mo rays is 1·43. We should expect the two ratios to be much the same. (See Table XXVIII, p. 159.)

Ionisation chambers are sometimes made of cylindrical form and shallow, the rays passing in a pencil down the axis of the cylinder. If the two ends of the chamber are of the same metal, this is practically equivalent to a

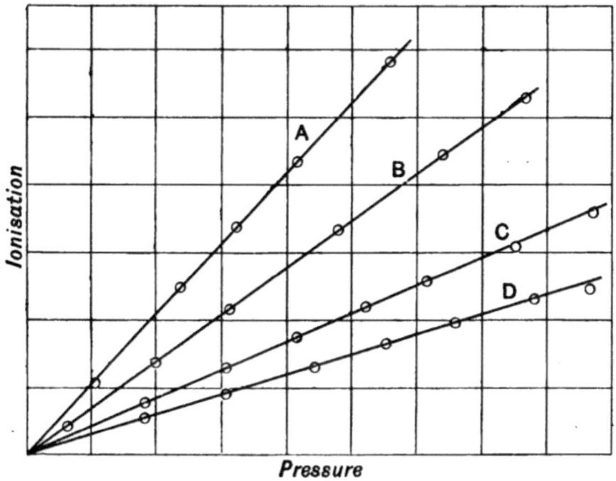

FIG. 60.

A. Ionisation in SeH_2—Sn rays. B. Ionisation in AsH_3—Sn rays.
C. ,, ,, SeH_2—Mo rays. D. ,, ,, AsH_3—Mo rays.

chamber of one material only. Laby and Kaye used a vessel of this kind to measure the variation of the ionisation due to an external source of γ rays with the pressure of the air inside the vessel (*Phil. Mag.*, December, 1908). The plates forming the ends of the chamber were of aluminium. The density of β radiation (*kd*, see p. 96) in aluminium is a little greater than in air, since the former substance has the larger atomic weight and the range of the β particle is somewhat greater therefore than in the air (see p. 98). The curves connecting ionisation with pressure should therefore bend over a little towards the pressure axis, and this was found to be the case (Fig. 61).

When the density of β radiation in the gas is greater than in the substance of the walls, the curve bends the other way. Examples of this are to be found in Beatty's paper already referred to. Fig. 62 shows the action of Sn X rays upon a shallow chamber 10 cm. in diameter and 1·6 cm. deep, the ends being made of paper; the gas was AsH_3.

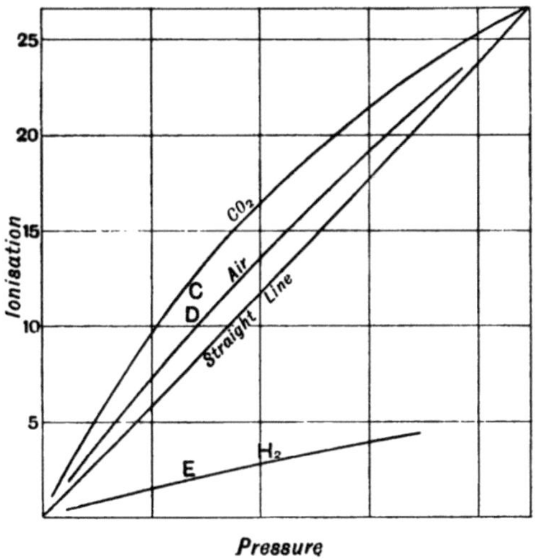

FIG. 61.—These curves show the relation between ionisation and pressure up to about 15 atmospheres. The curve for CO_2 is not drawn to quite the same scale as the others. For further details the original paper should be consulted.

The general form of such curves may be approximately found in the following way.

Let k and k' be the mass absorption coefficients of the material of the walls and of the gas for the X or the γ rays which are entering the chamber. Let λ and λ' be the absorption coefficients of the corresponding β rays in the same materials. Let us assume exponential laws provisionally, though no doubt this introduces error. Let

D be the product of the depth of the chamber and the density of the gas. The k coefficients are of course far smaller than the λ coefficients, and the depths of the layers in the walls which can furnish β rays to the chamber are so shallow that they absorb very little X radiation, and the variation of the X radiation in those layers can be neglected.

Consider two layers in the chamber walls of weight dx per sq. cm., one a distance x, measured by weight, above

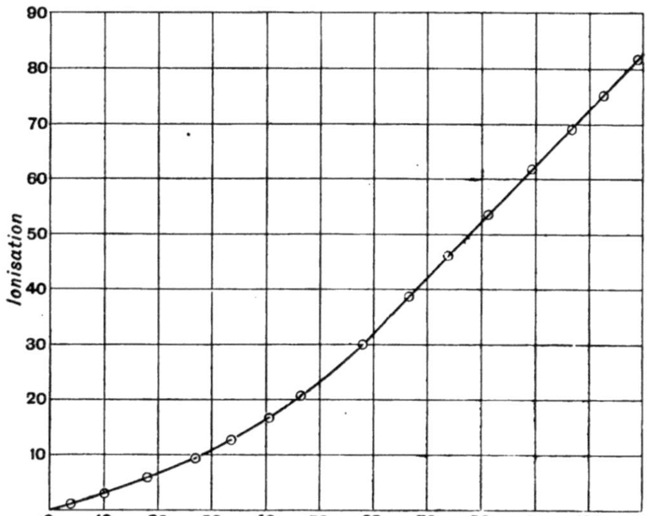

FIG. 62.—Action of Sn X-rays upon AsH_3 in a shallow ionisation chamber with paper walls.

the lower surface of the upper plate, the other a corresponding distance x below the upper surface of the lower plate. In each layer $Ikdx$ of the X corpuscles are converted into β rays, I being the whole number of corpuscles. Some in each layer move towards the chamber, and some away from it. The whole quantity moving towards the chamber is $Ikdx$. Of these we may assume that $Ikdxe^{-\lambda x}$ enter it, and in crossing the chamber spend a fraction of their energy equal to $1-e^{-\lambda' D}$. Integrating for all values of x, we find that the energy spent in the chamber by the β rays that came from the walls is equal to $Ik(1-e^{-\lambda' D})/\lambda$.

We have now to take account of the ionisation due to the β rays produced by the X rays in the gas in the chamber. If the walls and the gas were formed of the same atoms, the energy absorbed and spent on ionisation would be IDk', for the energy is spent and the ionisation is produced uniformly along the track of the X rays; distance along the track being estimated by the weight passed through, no matter whether the substance is solid or gaseous. If we substitute walls having constants k and λ for those of constants k' and λ' which are the constants of the gas, we add an expenditure of energy on ionisation $kI(1-e^{-\lambda' D})/\lambda$, but take away an amount $k'I(1-e^{-\lambda' D})/\lambda'$. Hence the whole ionisation in the vessel is represented by

FIG. 63.

$$I\left\{Dk' + \left(\frac{k}{\lambda} - \frac{k'}{\lambda'}\right)(1-e^{-\lambda' D})\right\}$$

When $k=k'$ and $\lambda=\lambda'$, this reduces to IDk'. When $\lambda' D$ is small, that is for low pressures of the gas, the expression becomes

$$I\left\{\frac{Dk\lambda'}{\lambda} + \left(\frac{k'}{\lambda'} - \frac{k}{\lambda}\right)\frac{\lambda'^2 D^2}{2}\right\}$$

and the curve is convex or concave to the pressure axis according as k'/λ' is greater or less than k/λ.

When Lenard's law holds, and we can assume that $\lambda=\lambda'$, the form of the curve then depends on the relative magnitudes of k and k'. The curve in Fig. 62 is an example.

When k/λ and k'/λ' are very different from each other, there can be no mistake as to the general form of the curve. In the case of γ rays the two quantities are, in general, more nearly equal than in the case of X rays, and the sign of $k'/\lambda' - k/\lambda$ will often depend on the relative magnitudes of λ and λ'.

There is an important consideration which we must not neglect when we apply this formula to the γ rays. The "returned" or "scattered" β radiation varies in quantity with the atomic weight of the substance on which it falls. In the calculation just made, we took no account of the scattering of the β rays that proceed from the upper wall of the chamber and strike the lower; and *vice versâ*. If cathode rays, that is to say β rays of the smaller velocities, are "absorbed" by matter according to a density law, we were perfectly justified in ignoring this effect, for there is then no change in the distribution of scattered rays when they pass from one medium to another. All matter acts in proportion to its weight, and geometrical spacing counts for nothing directly: the gas of the chamber and the material of the walls are continuations of each other for all purposes that relate to such β rays. But this does not hold for the swifter β rays, and the scattering or rather the change of scattering which takes place when the rays pass from one medium to another is not negligible.

We need not, however, enter into a calculation of the effects of scattering at the upper and lower walls, since we can look at the problem in a slightly different way. In a chamber of one material only, the density of the β rays is proportional to kd, where d has the meaning already given (p. 95). If the cavity in the material is filled with gas, having constants k' and d', the β ray density within it gradually changes from kd towards $k'd'$, attaining the latter value finally in such places in the gas as are out of reach of β rays from the walls. The extent of such places increases of course with the pressure of the gas. The formula should express this fact, and if it did, it would automatically take account of all scatterings.

When we attack the question in this way, the quantity d or some constant multiple of d naturally takes the place of $1/\lambda$, since the former gives the relative length of the track of the β ray apart from all scatterings. If, therefore, we make slight changes in the formula already obtained so as to adapt it to this view, it becomes

$$I\{Dk' + (kd - k'd')(1 - e^{-D'/d'})\}$$

If $kd = k'd'$ the ionisation is proportional to the pressure, as of course it should be. When D is small the expression becomes $ID.kd/d'$, and when D is large, $I\{Dk' + (kd - k'd')\}$.

The ratio of the final inclination to the inclination at the origin is $k'd'/kd$. This is clearly a correct representation, as, when the pressure is low and increasing, gas is being placed in regions where the β density is kd, while finally the accessions of gas may be considered as being placed in regions where the density is $k'd'$. If we assume, not only that the X rays spend no energy in the gas and in the chamber walls beyond that which goes to the production of cathode rays, but also that the absorption of cathode rays is proportional to the density of the material traversed, then the ratio of the final to the initial slope should be k/k'.

In the case of Beatty's curve (Fig. 62) for Sn X rays acting on AsH_3 between paper walls, these assumptions do not hold, because the X rays spend energy in producing secondary radiation from the arsenic. It is not, therefore, surprising that the ratio appears from the figure to be about 6, while the ratio k/k' is certainly much larger. We can say that d is greater than d'; and this is reasonable because the cathode ray in the gas sometimes spends its energy in making an As X ray, and so does not complete its full path. The average path is therefore shorter.

When the walls of the chamber or the gas itself are the source of homogeneous secondary radiation in sufficient quantity, the problem becomes much more complicated in general.

If the walls and the gas were composed of atoms of the same or nearly the same weight, and if the penetration of the secondary X rays were so much less than that of the primary that the absorption of the primary in the walls was negligible and the absorption of the secondary complete, the case would be exactly the same as the one already considered in respect to cathode rays alone. The density of the secondary X radiation as well as that of the cathode radiation would be the same at all points in the interior, and the density of the latter radiation would indeed be the same as if no secondary X radiation existed at all; for, if we may neglect the energies of transformation, the secondary X ray is only the occasion of a lull in the action of the cathode ray, stopping its action at one point, so to speak, only to transport it without loss of energy to some other point where it renews its career of ionisation.

The absorption coefficients of primary and secondary rays rarely differ by so much as to give these conditions.

If the walls of the chamber give off no appreciable secondary X radiation, the presence of such radiation having its origin in the gas does not affect the pressure-ionisation curve while the pressure is low. The curve in Fig. 62 will show a second bend when the secondary X ionisation fails to cross the chamber, just as a first bend occurs in the figure when the β-rays cannot cross it. But clearly this is a long way from the origin of the curve, and will not be a sharp one when it does come. At low pressures, of the fraction of the primary X rays which is absorbed in the gas, much less than half is transformed directly or indirectly into secondary X rays, the greater part becoming cathode radiation. Since the absorption coefficient of cathode radiation is of the order of one hundred times that of secondary *X* radiation, the latter contributes very little to the ionisation of the gas.

When the walls of the chamber give off secondary *X* rays in appreciable quantity and the gas does not, the action of the secondary rays upon the gas can be easily found experimentally; and if some law be found to connect this effect with the rate at which the secondary is produced in the material of the walls, the latter quantity can be determined in this way. We can then make an approximate calculation of the ionisation in a chamber of this sort in terms of the known absorption coefficients of the various rays, even when the conditions are not such as to permit the application of the simpler lines of argument just employed. The comparison of the calculated results with those experimentally found may be looked on as a test of the theory.

An attempt to carry out tests of this kind is described in a paper by Porter and myself (*Proc. Roy. Soc.* lxxxv, p. 349, May, 1911). It is unnecessary to repeat here the full description of the experiment and of the method of calculating the various quantities. A brief account of one such test will be sufficient.

Sn *X* rays are passed into an ionisation chamber, the walls of which can be made of different substances called radiators in Table XXXI. The chamber is cylindrical, and the rays are directed along the axis (see Fig. 49). The cathode radiations are measured as in the simpler cases described above; for example, by measuring the current where the upper wall is (*a*) bare to the chamber, (*b*) covered with a thin sheet of tissue paper, and subtracting the latter from the former. In order to find the effect due to secondary *X* radiation the various radiators, covered with tissue paper, are made to form in turn the upper wall of the chamber. Since no secondary *X* rays of appreciable importance are excited by Sn rays in aluminium the amount of secondary radiation in any other substance is easily determined by comparison.

For example, in an experiment in which the rays entered the chamber through a paper screen and fell upon an aluminium plate with the papered face down towards the chamber, the current was 136 in arbitrary units. When nickel with the papered face down was substituted for the aluminium the current was increased to 177, and if the nickel plate was turned over so as to present its bare side to the X rays and the chamber, the current became 296. Thus there was an incidence cathode radiation of 119 and a secondary X ray effect of 41. The former is best expressed in proportion to the effect of the X rays in crossing one cm. of air, and since the chamber is 4·2 cm. deep, we set it down as $4{\cdot}2 \times 119/136 = 3{\cdot}67$. The ratio of the emergence to the incidence cathode radiation was found to be 1·50 by separate experiment. Thus the whole ionisation due to the cathode radiation from the upper and lower walls, excited by a certain stream of Sn X rays crossing the chamber normally, is $3{\cdot}67 \times 2{\cdot}5 = 9{\cdot}18$ times the ionisation caused by these X rays in crossing one cm. of air.

We must now allow for the energy spent in making secondary X rays. Barkla has shown that such secondary rays are radiated uniformly in all directions about their origin. Let a pencil of X rays of energy I enter the upper plate of the ionisation chamber and let k_1 be the absorption coefficient of the plate for those rays. Let k_2 be the absorption coefficient of the same plate for its own characteristic rays. The loss of energy of the primary in crossing a stratum of weight dx is $k_1 I e^{-k_1 x} dx$, where x is the mass penetrated previously. Of this the amount that gets back into the ionisation chamber is

$$\tfrac{1}{2}\mu k_1 I e^{-k_1 x}\, dx \{e^{-k_2 x} + k_2 x\, Ei(-k_2 x)\}$$

an expression which contains an exponential integral.

(Soddy, *Phil. Mag.*, May, 1910, p. 735). The divisor 2 is introduced because half the energy may be taken to go forwards and half backwards : μ is the fraction of the primary energy which is directly or indirectly converted into secondary X ray energy, and the factor in brackets is necessary to allow for the absorption in the stratum of weight x when the rays are distributed in all directions ; if the rays had proceeded only by the shortest path across the stratum the amount absorbed would have been e^{-k_2x} and greater than the quantity in the bracket, since the exponential integral is negative.

The whole energy of secondary radiation which comes back and crosses the chamber is therefore

$$\tfrac{1}{2}\mu I\int_0^\infty k_1e^{-k_1x}\{e^{-k_2x}+k_2xEi(-k_2x)\}\,dx=\mu I\int_0^\infty \tfrac{1}{2}\frac{k_1}{k_2}e^{-\frac{k_1y}{k_2}}\{e^{-y}+yEi(-y)\}\,dy$$

and is a function of k_1/k_2 only.

The value of this integral can be found for a few values of k_1/k_2 by plotting a curve and finding the area. The result

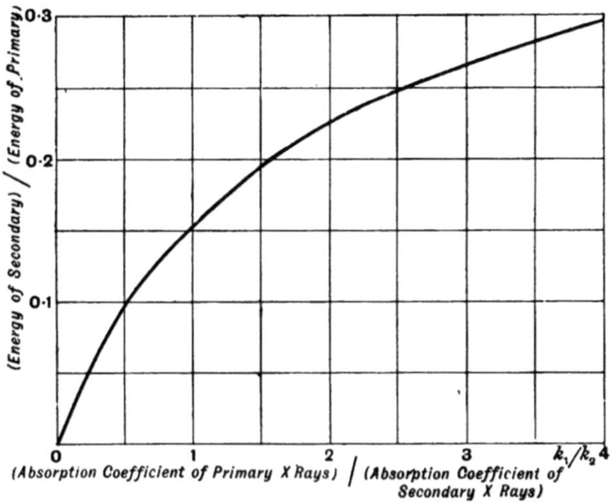

FIG. 64.

is shown in Fig. 64, where a line has been drawn through points representing the results of the calculations, and the curve gives the value of the integral sufficiently well for all values of k_1/k_2.

For example, when Sn rays are incident on copper, $k_1 = 14{\cdot}5$ and $k_2 = 53$. Thus $k_1/k_2 = 0{\cdot}274$ and the value of the integral is found from the curve to be 0·0588, or

$$\frac{\text{Energy of incidence secondary entering the chamber}}{\text{Energy of primary crossing the chamber}} = \mu \times 0{\cdot}0588.$$

It is perhaps worth observing that if the secondary rays moved only in the direction of the primary this ratio would be

$$\tfrac{1}{2}\mu \int_0^\infty k_1 e^{-k_1 x} e^{-k_2 x} dx = \frac{\mu k_1}{2(k_1 + k_2)} = \mu \times 0{\cdot}107.$$

Also, if the secondary rays moved only at right-angles to the primary, there would be no secondary X ray effect in the chamber if the primary rays met the walls normally. The actual ratio we have obtained, viz. $\mu \times 0{\cdot}0588$, lies between $\mu \times 0{\cdot}107$ and zero.

To return to the argument, the absorption coefficient of Sn rays in Al is 1·83 nearly, and of Cu rays in Al is 47·7. The ratio of the absorption coefficients of these radiations in air is nearly the same as in Al, [1] and the ionisation produced is in proportion to the absorption, hence :—

$$\begin{aligned}\frac{\text{Ionisation in chamber due to secondary rays}}{\text{Ionisation in chamber due to primary rays}} &= \mu \times 0{\cdot}0588 \times 47{\cdot}7/1{\cdot}83. \\ &= \mu \times 1{\cdot}53\end{aligned}$$

This assumes that the secondary rays have the same length of track in the chamber as the primary rays, and the apparatus was designed to make this the case as nearly as possible; the condition can only be satisfied approximately. See Fig. 54.

[1] I have here used absorption coefficients in Al rather than in C, as in the original paper. As Prof. Barkla has pointed out (*Phil. Mag.*, Feb. 1912), the absorption coefficient of the Sn X rays in C includes an important scattering coefficient, and is therefore somewhat uncertain. The change does not make any material difference to the main conclusions to be drawn from Table XXXI.

By actual experiment the ratio of these ionisations is found to be 0·301 and therefore $\mu = 0{\cdot}197$.

We may say, therefore, that when Sn rays pass through copper 0·197 of the energy absorbed is spent in making Cu *X* rays and only the remainder in making cathode rays; and so if we measure the cathode radiations that emerge into the chamber and thence calculate—if we can—their original amount, we ought to find the latter to be proportional to that remainder. This reasoning is not accurate, because our knowledge of the energy changes and of the whole process generally is far from perfect. In particular, as I have already said, there is some doubt as to the absorption coefficient of the cathode rays in all cases, and it is not a simple matter therefore to find the original cathode radiation from that which is observed to enter the chamber. Nevertheless, the attempt seems quite worth while and Table XXXI. shows the result. The last column shows the ratio between the fraction of the absorbed Sn *X* rays which may be considered as the origin of the cathode rays and the quantity of cathode radiation which is found.

A word of explanation of column VII is required. When cathode radiations are measured in the manner described, the quantity found is the difference between the cathode radiation of the radiator and the cathode radiation from tissue paper. It is therefore comparable with the difference between the corresponding figure in column VI and the absorption coefficient of tissue paper; and 0·21, which is approximately the absorption coefficient of Sn rays in C, is subtracted from the figure in column VI to give the figure in column VII. This correction only effects in reality the result for aluminium.

On the whole, the figures in the last column agree very well; and the irregularities occur where we are most ignorant of certain data underlying the calculation, that is to say, the absorption coefficients of the β ray.

TABLE XXXI.—Sn Rays.

I. Radiator.	II. Cathode rays.	III. Secondary X rays.	IV. μ.	V. k.	VI. $k(1-\mu)$.	VII. $k(1-\mu)-0{\cdot}21$.	VIII. VII/II.
Al . . .	1·30	0	0	1·83	1·83	1·62	1·25
Fe . . .	9·02	0·184	0·098	14·3	13·0	12·8	1·42
Ni . . .	9·18	0·301	0·139	17·1	14·7	14·5	1·58
Cu . . .	10·26	0·301	0·197	17·4	14·0	13·8	1·34
Zn . . .	10·72	0·308	0·212	17·4	13·7	13·5	1·26
Sn . . .	12·36	0	0	12·9	12·9	12·7	1·03

Column I.—Name of metal in which the cathode and secondary X rays are excited.
,, II.—Amount of cathode radiation measured as described above.
,, III.—Amount of secondary X radiation.
,, IV.—Calculated value of ratio of secondary to primary X-ray energy.
,, V.—Absorption coefficient of primary X rays by the metal.
,, VI.—Proportion of the absorption which should go to the production of cathode rays.
,, VII.—The same, less the absorption coefficient of paper (see above).
,, VIII.—Ratio of VII to II.

The following table gives the results of similar experiments in which As X rays were used :—

TABLE XXXII.—As Rays.

I. Radiator.	II. Incidence cathode rays.	III. Secondary X rays.	IV. μ.	V. k.	VI. $k(-\mu)$.	VII. $k(1-\mu)-2{\cdot}5$.	VIII. VII/II.
Al . . .	0·115	0	0	23·6	23·6	21·1	183
Fe . . .	0·420	0·187	0·200	127	101·6	99·1	235
Ni . . .	0·500	0·165	0·293	147	106	103·5	207
Cu . . .	0·483	0·217	0·340	155	102	99·5	206
Zn . . .	0·536	0·190	0·371	186	117	114·5	213
Sn . . .	0·575	0	0	112	112	109·5	191

For description, see notes to Table XXXI.

CHAPTER XVI

THE SCATTERING OF X AND γ RAYS

As an X or γ ray moves on its way there is, as we have seen, a chance that it may disappear in crossing some atom and be replaced by a β ray, and the chance that this will happen in crossing a layer of any substance weighing one gramme to the square centimetre is the mass absorption coefficient in that substance. This is not the only event, however, that may take place during the motion. There is a small chance, much smaller than the other, that it may be deflected. That is to say, a small number of rays of the same quality as the primary are found proceeding in various directions from the matter struck, their distribution in space bearing some relation to the original direction of the primary stream. A complete determination of the whole effect would include measurements of the intensity of the radiation at every inclination to the direction of the primary, for rays of all qualities and atoms of all weights. This has by no means been achieved, but results have been obtained which are fairly representative of all the possible variations.

It is known, for example, that the scattered γ radiation springing from a screen through which the primary rays are passing normally is very much greater in quantity on the emergence than on the incidence side. The screen must not be thick enough to absorb a large part of the primary stream, or the comparison would, of course, be unfair. The same sort of dissymmetry is to be found in the case of the scattered γ rays as we considered in Chapter VIII. in respect to the secondary β rays.

An investigation of the dissymmetry was made by Madsen (*Trans. Roy. Soc. of South Australia*, Oct., 1908), who used an arrangement of apparatus very similar to that which served to compare the incidence and emergence secondary β rays (see p. 116). He estimated the emergence γ rays to be 4·5 to 6·5 times as great as the incidence. That is to say, when the stream of γ rays passed through thin plates of carbon, aluminium, zinc, or lead, five or six times as many scattered γ rays continued to some extent the motion of the primary rays as turned back and came out of the plate on the incidence side. Madsen also found a difference in quality on the two sides, which may seem to clash with the definition of scattered radiations given above, but it is to be remembered that the differences in quality may be due to want of uniformity in the primary stream. No experiments have been made with γ rays known to be homogeneous. Madsen states that "there appears to be reason to believe that the distribution of the scattered radiation depends to some extent upon the hardness of the radiation which is scattered; also upon the nature of the material in which the scattering is produced. The softer radiation appears to be turned back to a somewhat greater extent than the hard. Materials of high atomic weight seem to be able to produce more complete scattering than those of low atomic weight."

The resemblance to the scattering of β rays is very striking.

The subject has been more fully investigated by Florance (*Phil. Mag.*, Dec. 1910) who was able to employ large quantities of radioactive material and therefore to obtain more accurate results. The general arrangement of the apparatus is represented in Fig. 65.

The electroscope could be moved into various positions as

Fig. 65 shows, and the amount of γ radiation scattered in various directions could therefore be determined. The arrangement was suitably varied when it was desired to measure the incidence secondary rays. Fig. 66 is a reproduction of the diagram in which Florance exhibits his results, the radius vector in any direction being a measure of the radiation scattered in that direction. Madsen's experiments dealt only with the whole quantities of the

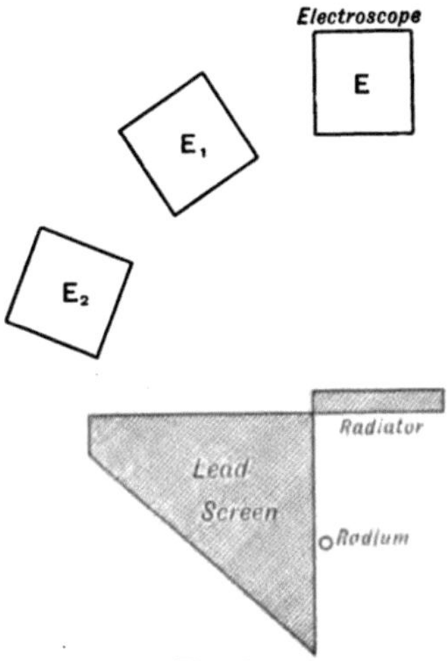

FIG. 65.

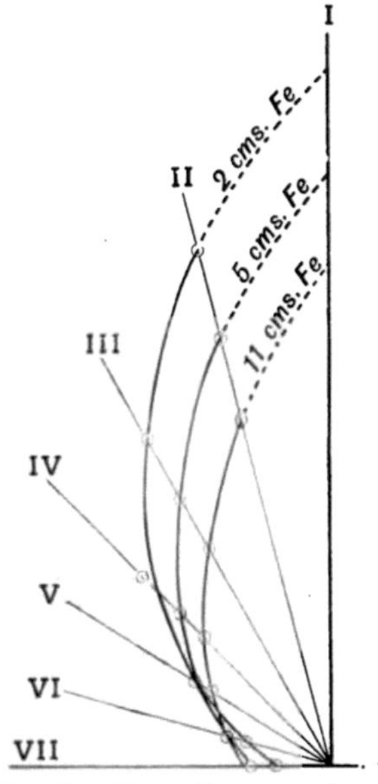

FIG. 66.—The radiators are iron plates of various thicknesses.

emergence and incidence radiations; these go much more into detail. They confirm Madsen's general conclusions with some little modification, and extend them considerably, as will be evident from the following extracts taken from Florance's summary.

"The 'incident' secondary is in all cases softer than the 'emergent' secondary. There is, moreover, a gradual change from the quality of the primary to that of the secondary emergent and then to that of the secondary incident. The quality therefore depends on the position of the electroscope." That is to say, the penetration of the rays diminishes as we proceed from I to VII in Fig. 66 and thence to the incidence rays. Florance states that the linear coefficient of absorption by lead of the primary rays was 0·68 ; of those scattered at an angle of 25°, 1·20 ; and for an angle of 55°, 1·77.

"There is a gradual change in the quantity of secondary γ radiation from that which emerges from the radiator in the direction of the original radiation to that which is returned in the reverse direction."

Florance finds that the secondary radiation, like the primary, is heterogeneous, and that the quality can vary with the form and the nature of the radiator; but he concludes that these variations are entirely due to differences in the scattering of the harder and softer constituents of the primary radiation, and that the character of the secondary γ radiation due to a homogeneous primary is independent of the nature of the atom to which it is due.

Thus, under closer examination, the parallelism between the scattering of β rays and that of γ rays continues to be very strongly marked.

Investigations of the scattering of X radiation have led to results which are in most respects parallel to those first outlined.

Bragg and Glasson have shown (*Trans. Roy. Soc. of South Australia*, October, 1908, *Phil. Mag.*, June, 1909) that when a pencil of primary X rays passes normally across a thin sheet of matter, the emergence secondary X radiation is greater than the incidence, except when the secondary radiation consists of the characteristic radiation of the material of which the sheet is composed. The arrangement of the experiment is shown in Fig. 67.

A narrow pencil of X rays passes upwards through

apertures in lead plates at A and B, along the axis of the ionisation chamber, and out into the open. The central part of the apparatus in the figure consists of a double cylinder of brass; a cylinder 2 in. long, and 4 in. in diameter is connected to a cylinder 2 in. long, and 2 in. in diameter by a connecting piece DD, shown in Fig. 67 ii. The latter resembles a light brass wheel with four spokes,

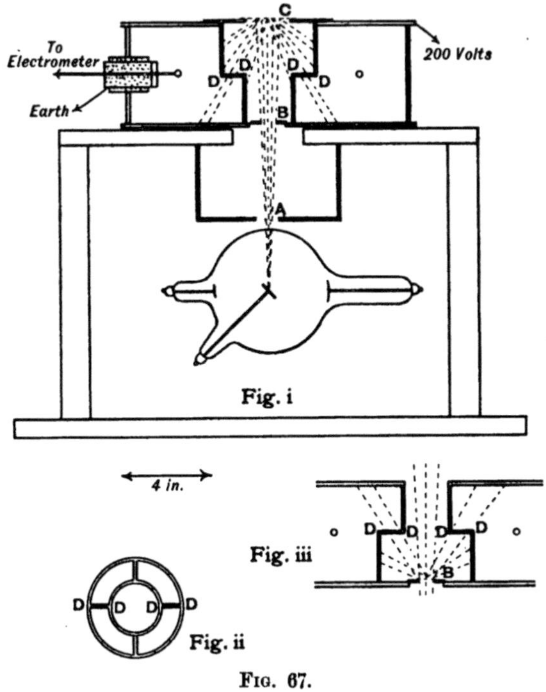

FIG. 67.

and various thin screens in the form of flat rings can be attached to it, filling up the spaces between the spokes. In Fig. 67 i the double cylinder is shown as arranged for the measurement of incidence secondary radiations. The radiating sheet lies at C, and is struck by the pencil of X rays; the secondary incidence rays strike downwards, and some pass through the screen on DD into the ionisation chamber as shown. Those that strike the walls of the brass cylinder are stopped. The current is also measured when the screen is removed from *C*, and the difference between the two observations is taken as a measure of the incidence *X* rays. The double cylinder is then inverted as shown in Fig. 67 iii, and the screen now placed at *B*. The difference between the currents observed before and after putting the screen at *B* is taken as a measure of the emergence secondary rays. In this way the pencils of secondary rays are confined to limits of about 30° and 50° for the emergence, and 130° and 150° for the incidence, the angles being those made with the axis of the cylinder.

For example, when the radiator at *C* was of aluminium 0·4 mm. thick, and the absorbing screen *DD* of tinfoil (weighing 0·0056 gr. per sq. cm.) the currents with and without the radiator at *C* in Fig. 67 i were 86 and 26 respectively, in arbitrary units; the currents with and without the same screen at *B* in Fig. 67 iii were 220 and 35 respectively. The incidence and emergence radiations were, therefore, 60 and 185, and the latter was three times as large as the former. Other experiments of the same kind are described in the paper.

More detailed examinations of the distribution of the scattered radiation have been made by Barkla (*Phil. Mag.*, Feb., 1911) and Crowther (*Proc. Roy. Soc.*, lxxxv Jan., 1911, p. 29).

The former directed a pencil of primary *X* rays upon a carbon plate as in Fig. 68. The scattered rays caused ionisation in two electroscopes *A* and *A'*; of which the former was so placed that it received scattered rays making an angle of 90° with the primary pencil, and the latter received scattered rays making the same angle with the plate as was made by the rays affecting *A*. In this way it was ensured that the two bundles of scattered rays were equally absorbed during their exit from the plate.

The determination of the two currents gave the ratio of

the amounts scattered in the direction at right angles to the primary rays, and in any other given direction. To determine the intensity of the scattered radiation in directions making small angles with the direction of the primary rays, it was necessary to work with the scattered rays emerging from the other side of the plate.

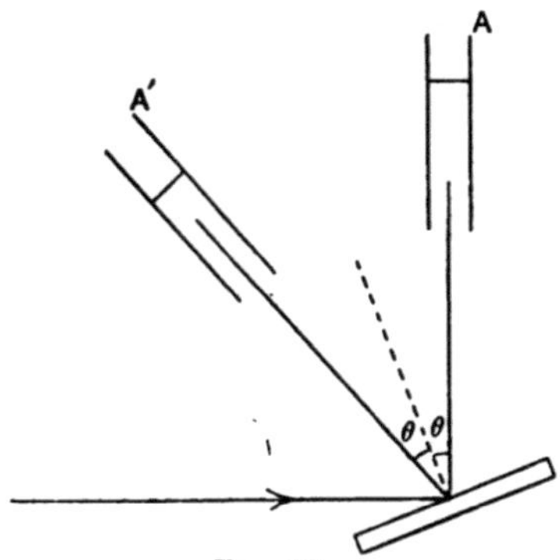

Fig. 68.

Barkla's results are plotted (dots in circles) in Fig. 69, where OA is the direction of propagation of the primary radiation and the length of each radius vector represents the intensity of radiation in the direction in which it is drawn.

The two points marked P and Q are not placed by Barkla on the diagram from which this figure is drawn.

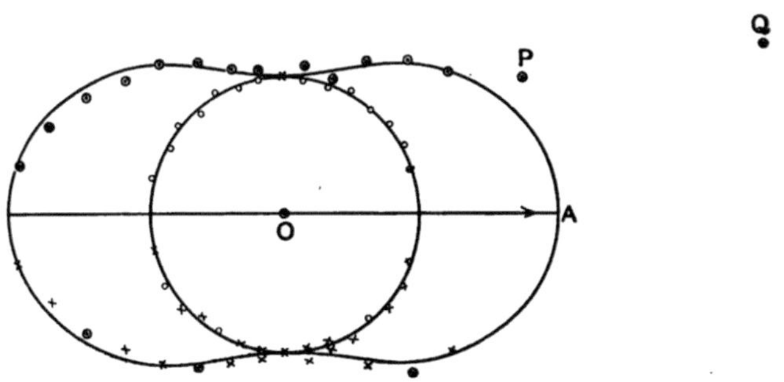

Fig. 69 (Barkla).—The dots in circles show the results of the experiments referred to in the text; the small circles, the results when a radiator giving homogeneous secondary rays was substituted for the carbon plate. The crosses represent results assumed from symmetry. The finer lines are calculated forms.

They represent results which he obtained, but he did not consider they should be included, for, as he says, "in these cases the correction for scattering from air was large and the results were much less reliable. The fact that the discrepancy appears only for these small angles is significant. It seems just possible that these results are vitiated by an irregular refraction effect." In no other case has the refraction of X-rays been proved, and the presence of dissymmetry in scattering has been clearly shown with all the new rays, so that it seems right to put these results with the rest which Barkla obtained.

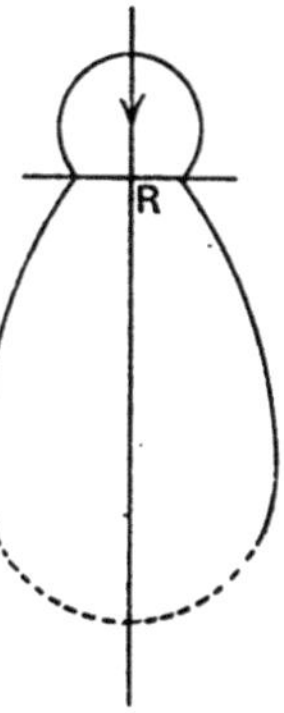

FIG. 70.

Crowther's arrangement was a little different from that of Barkla in that the radiator was so thin that the correction for absorption was small and it was considered unnecessary to adjust the position of the radiator for each fresh experiment, as Barkla's method required. The results of Crowther's observations are given in Fig. 70, which is to be compared with Fig. 69.

It will be seen that Crowther obtains a marked dissymmetry which is in agreement with the result found by Glasson and myself. His distribution of the scattered rays does not accord well with that of Barkla; the emergence side shows a curve resembling more closely that of Florance for scattered γ rays, Fig. 66.

The re-entrant angle on each side of the curve, more marked in Crowther's than in Barkla's, is a notable feature and will be considered later.

CHAPTER XVII

THE NATURE OF THE X AND γ RAYS

In the preceding chapters I have tried to show that the X and γ rays must be considered to be corpuscular. I have adopted a definition of this latter term which does not bring in the word material, my purpose being to avoid limitations which might prove unnecessary and misleading.

The question now arises as to whether greater precision can be given to the definition, and the rays linked more closely to other known phenomena and to proved theories.

The main properties for which we have to account are the curious mutual interchangeability between the X ray and the electron, the electrical neutrality of the X ray, and the polarisation already referred to. If Marx's experiment is right, we must also explain why the X rays travel with the velocity of light, and, further, a complete theory must lead to the observed laws of scattering and absorption.

The most famous theory of the X ray is that proposed by Sir George Stokes. When an electron is accelerated in any way energy is radiated from the place of acceleration through the æther in what may be called an æther pulse. Such a disturbance, if thin enough, will have the negative qualities of the X ray: it will be incapable of reflection, refraction, and polarisation as affected by crystalline structure; and diffraction effects will be beyond observation. It will have the positive property of moving with the velocity of light. If secondary *X* rays are assumed to be disturbances of the æther arising from accelerations of the electrons in the atoms swept over by primary *X* rays, then the polarisation which Barkla found is qualitatively explained, and with this goes the existence of the nicks in the curves of Figs. 69 and 70 (Barkla, *Phil. Mag.*, February, 1911, p. 270). These last are striking agreements between theory and experiment.

But beyond this point the theory does not seem to make satisfactory progress. It may well be supposed that the failure is due to the fundamental defect that it cannot explain the interchangeability of the *X* ray and the electron. It cannot show how the *X* ray carries away so large a fraction (possibly the whole) of the energy of one electron and hands it over to another. If the theory cannot express this chief result of experiment, if indeed it tends to hide and ignore it, we cannot wonder at its lack of power as a further guide to experimental research. The most striking quantitative results are connected with the handing of energy from the *X* ray to the electron, and back again. But apart from these the assumptions made in respect to the origin of the *X* rays lead to deductions concerning their power of penetrating materials (J. J. Thomson, "Conduction of Elect. through Gases," Art. 162) which are not to be reconciled with experiment except by various further assumptions of a very special nature. In other words. the experiments give no support to the theory.

Much the same can be said in respect to the calculations of the scattering of the *X* ray, for although the calculated form of the scattering curve, Fig. 69, does fit the experimental curve in some parts, there are wide differences in others. The pulse theory gives no explanation of the dissymmetry between the rays scattered forwards and backwards, a dissymmetry which is so great in the case of the γ rays. Nor does it explain the dissymmetry in the ejection of the secondary cathode or β rays. It is sometimes said that the dissymmetry is due to the fact that the pulse has momentum to hand on, but this explanation is hopelessly insufficient until the pulse can be shown to be concentrated in a very small volume which does not spread as it travels; that is to say, until the fundamental point of interchangeability is mastered. There is a dissymmetry in the distribution of the X rays produced by cathode rays which Sommerfeld has lately discussed on the pulse theory (*Bayer, Akad. der. Wiss.* January 7, 1911). He shows that when an electron is brought to a speed of 99 per cent. of that of light, the disturbance travels outwards in a sort of hollow cone of 10° vertical angle, the axis of the cone being the direction of motion of the electron. When the final speed is 90 per cent. of that of light the angle is 50°, and so on. But this is as far as ever from explaining the interchange.

It is worth while referring to the point of the relative energies of the β rays and γ rays, since this may have a bearing on the choice of theories. If the γ rays are supposed to be due to pulses arising from the expulsion of β rays, the energy of the former must be less than that of the latter and in general considerably less (Sommerfeld, *loc. cit.*, p. 24). There should also be a connection between the energies of the two which is independent of the nature of materials involved. On a corpuscular theory, the γ may equally well be looked on as the original and the β as the secondary ray; no connection between the energies of the two kinds of ray can be foretold in the absence of knowledge as to how the radiation takes place. Probably the ratio would also depend on the nature of materials in the same way that it does in any stream of γ radiation. In the case of the rays from RaC, Eve has recently found the energy of the γ rays to be about twice as great as that of the β rays (*Phil. Mag.*, Oct., 1911, p. 551).

In the early days of X ray discovery, the pulse theory had some success in furnishing qualitative explanations. But, surely, it has made very little progress since that day and instead of leading, has rather lagged behind the general advance. The reason is that it delivers no attack on the central position, which is, as I have already said, the interchangeability of electron and X ray. Clinging to its old base it is, perhaps only for the time, unable to do so. It is necessary to adopt a new base if only to avoid stagnation, and we must seek that one from which attack will be most direct. Let us forget for the time that idea of keeping touch with electromagnetic theory as we fancy it must be, which is hampering every movement.

If we try to construct a theory which shall make the explanation of the interchangeability its principal feature, we are first led to conceive of a more material X ray. The electron of the β ray may be imagined as capable of attaching to itself enough positive electricity to neutralise its own charge and of doing this without appreciable addition to its mass. This is the transformation from electron to X ray : the reversed transformation occurs when the electron puts down its positive again. Neither change can occur, except during the passage of the entity through an atom. As an electron, the entity is capable of ionising and so forth, and it has little power of penetration since it easily loses energy. As an X ray, the entity, being neutral, passes through atoms freely and carries its store of energy from point to point without loss. When the X ray is scattered, the whole entity is swung off in a new direction.

It is no argument against this view that the positive electron has not yet been isolated, for the possibility of detecting a charged particle depends on the ratio of its charge to its mass. We can distinguish the charged atom, and the electron with an "e/m" ratio a thousand times greater than that of the atom; but it does not follow that we should as easily find a particle for which the ratio is much greater still. Nor is it an insuperable objection that the polarisation of the X ray does not find so ready an explanation as can be given on the pulse theory; nor, again, that the velocity of the X ray may be equal to that of light. A hypothesis is not to be set aside because it does not supply an immediate explanation of every fact; moreover, this particular hypothesis is by no means essentially incapable of meeting either of these objections.

The great bulk of the X ray phenomena are just what we should expect if we thought the electron able to neutralise its electric charge without alterations of any other of its properties or qualities. The neutral pair theory is a direct physical expression of the fact. It succeeds therefore exactly where the pulse theory fails, giving a simple and convenient means of picturing the X ray processes to the mind. To make the pulse theory a success, or perhaps it should be put, to fit the X ray into a scheme of electromagnetic radiation, it must be shown that the existence of a quantum behaving like a neutral pair can be reconciled with the laws of electromagnetism and is an extreme case of that which we know from another point of view as a wave of light. I think this has not yet been done. When and if it is accomplished the neutral pair idea will not have been thrown away, for it expresses a number of facts too simply and naturally; it will rather have been built into some greater structure.

Einstein, Stark, and others have been led to postulate a light-quantum; and in the photo-electric effect they see a transference of energy from the quantum to the electron. When I first put forward the neutral pair theory I was ignorant of the work of Einstein and was guided only by the results of experimental investigation on the behaviour of the new rays. I did not think of carrying over the idea to the theory of light; on the contrary, I had hopes of proving that no connection existed between the two kinds of radiation. It still seems to me that the neutral pair theory correctly pictures the chief processes of the *X* ray, which the old form of spreading pulse, even the modified Thomson's pulse, are unable to do. But I should now add that we ought to search for a possible scheme of greater comprehensiveness, under which the light wave and the corpuscular *X* ray may appear as the extreme presentments of some general effect.

To do this, the extreme views should be applied to all the phenomena of both light and *X* rays in order to find out how far each can be made effective. As regards the application of the electromagnetic theory—which fits light effects so well—to the phenomena of the *X* ray, a great deal of work has been done and we know its strength and its weakness. Very little has been done in the converse direction. The interchangeability which the neutral pair theory expresses is abundantly illustrated in the behaviour of the *X* ray. It will be very interesting, I think, to carry over the ideas which we learn in this part of the field to that other part where we consider the relation between electron movement and radiation through the æther. The *X* ray phenomena suggest to us that an electron of given energy may be converted into a light-quantum of equal energy and *vice versa*, that the chance of either conversion is a function of the energy and depends also on the nature of the material which is required to effect the conversion, and that, in consequence, radiation of a certain composition must exist in equilibrium with a given form of electron movement such as the thermal agitation of electrons in a metal. If investigation from this point of view proves successful, we shall I think be guided and spurred on towards some great idea which will reconcile the old antagonism between the corpuscle and the wave.